Water Pollution and Its Control

Water Pollution and Its Control

Anil Goel

MD PUBLICATIONS PVT LTD
New Delhi
www.mdppl.com

Published by
MD Publications Pvt Ltd
"MD House", 11, Darya Ganj,
New Delhi-110002
Ph. : +91-11-41563325, 41562846
E-mail : **contact@mdppl.com**
Website : **www.mdppl.com**

ISBN : 978-81-7533-128-0

Published and Printed by
MD Publications Pvt Ltd at Printext,
D2/149A, Jeewan Park, Uttam Nagar, New Delhi.

PREFACE

A change in the chemical, physical, biological and radiological quality of water which is injurious to its existing, intended, or potential uses is termed as "water pollution". The term "Water Pollution" generally refers to human-induced changes to water quality. Thus, the discharge of toxic chemicals from a pipe or the release of livestock waste into a nearby water body is considered pollution. Conversely, nutrients that originate from animals in the wild or toxins that originate from natural processes are not considered pollution.

The contamination of ground water, rivers, lakes, wetlands, estuaries, and oceans can threaten the health of humans and aquatic life. Sources of water pollution are generally divided into two categories. The first is point-source pollution, in which contaminants are discharged from a discrete location. Sewage outfalls and oil spills are examples of point-source pollution. The second category is non-point-source or diffuse pollution, referring to all of the other discharges that deliver contaminants to water bodies. Acid rain and unconfined runoff from agricultural or urban areas are examples of non-point-source pollution. The principal contaminants of water include toxic chemicals, nutrients and biodegradable organics, and bacterial and viral pathogens.

Water pollution can threaten human health when pollutants enter the body via skin exposure or through the direct consumption of contaminated food or drinking water. Priority pollutants, including dichlorodiphenyl trichloroethane (DDT) and polychlorinated biphenyls (PCBs), persist in the natural environment and bioaccumulate in the tissues of aquatic organisms. These persistent organic pollutants are transferred up the food chain in a process called biomagnification and they can reach levels of concern in fish species that are eaten by humans. Finally, bacteria and viral pathogens can pose a public health risk for those who drink contaminated water or eat raw shellfish from polluted water bodies.

Contaminants have a significant impact on aquatic ecosystems. for example, enrichment of water bodies with nutrients principally nitrogen and phosphorus can result in the growth of algae and other aquatic plants that shade or clog streams. If wastewater containing biodegradable organic matter is discharged into a stream with inadequate dissolved oxygen, the water downstream of the point of discharge will become anaerobic and will be turbid and dark. Over the reach of stream where the dissolved-oxygen concentration is zero, a zone of putrefaction will occur with the production of hydrogen sulfide, ammonia, and other odorous gases.

Wastewater discharges are most commonly controlled through effluent standards and discharge permits. Under this system, discharge permits are issued with limits on the quantity and quality of effluents. Water-quality standards are sets of qualitative and quantitative criteria designed to maintain or enhance the quality of receiving waters. Receiving waters are divided into several classes depending on their uses, existing or intended, with different sets of criteria designed to protect uses such as drinking water supply, bathing, boating, fresh-water and shellfish harvesting, and outdoor sports for seawater.

This book discusses the characteristics of groundwater and its contamination. It includes the types of aquifers, the factors affecting the flow of groundwater, and its natural characteristics. It describes how groundwater is contaminated and the types of contamination most commonly found. What is known about the effects of various types of contaminants on human health and the environment are briefly discussed.

I am indebted to the various writers from whose works I have drawn inspiration and have quoted to bolster the arguments postulated in the book. I am also grateful to my family for bearing with me during the research and editing of this book and to Mr. Pranav Gupta of M/s. MD Publications Pvt. Ltd., Delhi for having brought out the book well and in such a short time.

Anil Goel

CONTENTS

1

Water Pollution: Definition and Meaning

Introduction

Water pollution is a large set of adverse effects upon water bodies such as lakes, rivers, oceans, and groundwater caused by human activities. Although natural phenomena such as volcanoes, algae blooms, storms, and earthquakes also cause major changes in water quality and the ecological status of water, these are not deemed to be pollution. Water pollution has many causes and characteristics. Increases in nutrient loading may lead to eutrophication. Organic wastes such as sewage impose high oxygen demands on the receiving water leading to oxygen depletion with potentially severe impacts on the whole eco-system. Industries discharge a variety of pollutants in their wastewater including heavy metals, organic toxins, oils, nutrients, and solids. Discharges can also have thermal effects, especially those from power stations, and these too reduce the available oxygen. Silt-bearing runoff from many activities including construction sites, deforestation and agriculture can inhibit the penetration of sunlight through the water column, restricting photosynthesis and causing blanketing of the lake or river bed, in turn damaging ecological systems.

Pollutants in water include a wide spectrum of chemicals, pathogens, and physical chemistry or sensory changes. Many of the chemical substances are toxic. Pathogens can obviously produce waterborne diseases in either human or animal hosts. Alteration of water's physical chemistry include acidity, conductivity, temperature, and eutrophication. Eutrophication is the fertilisation of surface water by nutrients that were previously scarce. Even many of the municipal water supplies in developed countries can present health risks. Water pollution is a major problem in the global context. It has been suggested that it is the leading worldwide cause of deaths and diseases, and that it accounts for the deaths of more than 14,000 people daily.

Sources of Water Pollution

Some of the principal sources of water pollution are:

- geology of aquifers from which groundwater is abstracted
- industrial discharge of chemical wastes and byproducts
- discharge of poorly-treated or untreated sewage
- surface runoff containing pesticides or fertilizers
- slash and burn farming practice, which is often an element within shifting cultivation agricultural systems
- surface runoff containing spilled petroleum products
- surface runoff from construction sites, farms, or paved and other impervious surfaces e.g. silt
- discharge of contaminated and/or heated water used for industrial processes
- acid rain caused by industrial discharge of sulfur dioxide (by burning high-sulfur fossil fuels)
- excess nutrients added (eutrophication) by runoff containing detergents or fertilizers
- underground storage tank leakage, leading to soil contamination, thence aquifer contamination

Accumulation of litter and organic debris in Salford Quays, a section of the Manchester Ship Canal in Greater Manchester, UK. Raw sewage and industrial waste flows into the U.S. from Mexico as the New River passes from Mexicali, Baja California to Calexico, California.

Contaminants

Contaminants may include organic and inorganic substances. Some organic water pollutants are:

- insecticides and herbicides, a huge range of organohalide and other chemicals
- bacteria, often is from sewage or livestock operations;
- food processing waste, including pathogens
- tree and brush debris from logging operations
- VOCs (Volatile Organic Compounds, industrial solvents) from improper storage

Some inorganic water pollutants include:

- heavy metals including acid mine drainage
- acidity caused by industrial discharges (especially sulfur dioxide from power plants)
- chemical waste as industrial by products
- fertilizers, in runoff from agriculture including nitrates and phosphates
- silt in surface runoff from construction sites, logging, slash and burn practices or land clearing sites

Transport and Chemical Reactions of Water Pollutants

Most water pollutants are eventually carried by the rivers into the oceans. In some areas of the world the influence can be traced hundred miles from the mouth by studies using hydrology transport models. Advanced computer models such as SWMM or the DSSAM Model have been used in many locations worldwide to examine the fate of pollutants in aquatic systems. Indicator filter feeding species such as copepods have also been used to study pollutant fates in the New York Bight, for example. The highest toxin loads are not directly at the mouth of the Hudson River, but 100 kilometers south, since several days are required for incorporation into planktonic tissue. The Hudson discharge flows south along the coast due to coriolis force. Further south then are areas of oxygen depletion, caused by chemicals using up oxygen and by algae blooms, caused by excess nutrients from algal cell death and decomposition. Fish and shellfish kills have been reported, because toxins climb the foodchain after small fish consume copepods, then large fish eat smaller fish, etc.

Each successive step up the food chain causes a stepwise concentration of pollutants such as heavy metals (e.g. mercury) and persistent organic pollutants such as DDT.

The big gyres in the oceans trap floating plastic debris. The North Pacific Gyre for example has collected the so-called *Great Pacific Garbage Patch* that is now about the size of Texas. Many of these long-lasting pieces wind up in the stomachs of marine birds and animals. Many chemicals undergo reactive decay or change especially over long periods of time in groundwater reservoirs. A noteworthy class of such chemicals are the chlorinated hydrocarbons such as trichloroethylene (used in industrial metal degreasing) and tetrachloroethylene used in the dry cleaning industry. Both of these chemicals, which are carcinogens themselves, undergo partial decomposition reactions, leading to new hazardous chemicals.

Groundwater pollution is much more difficult to abate than surface pollution because groundwater can move great distances through unseen aquifers. Non-porous aquifers such as clays partially purify water of bacteria by simple filtration (adsorption and absorption), dilution, and, in some cases, chemical reactions and biological activity: however, in some cases, the pollutants merely transform to soil contaminants. Groundwater that moves through cracks and caverns is not filtered and can be transported as easily as surface water. In fact, this can be aggravated by the human tendency to use natural sinkholes as dumps in areas of Karst topography.

There are a variety of secondary effects stemming not from the original pollutant, but a derivative condition. Some of these secondary impacts are:

- Silt bearing surface runoff from can inhibit the penetration of sunlight through the water column, hampering Photosynthesis in aquatic plants.
- Thermal pollution can induce fish kills and invasion by new thermophyllic species freaking

These harmful materials included organic wastes, sediments, minerals, nutrients, thermal pollutants, toxic chemicals, and other hazardous substances. Organic wastes are produced by animals and humans, and include such things as fecal matter, crop debris, yard clippings, food wastes, rubber, plastic, wood, and disposable diapers. Such wastes require oxygen to decompose. When they are dumped into streams and lakes and begin to break down, they can deprive aquatic life of the oxygen it needs to survive.

Sediments may be deposited into lakes and streams through soil erosion caused by the clearing, excavating, grading, transporting, and filling of land. Minerals, such as iron, copper, chromium, platinum, nickel, zinc, and tin, can be discharged into streams and lakes as a result of various mining activities. Excessive levels of sediments and minerals in water can inhibit the penetration of sunlight, which reduces the production of photosynthetic organisms.

Nutrients, like phosphorus and nitrogen, support the growth of algae and other plants forming the lower levels of the food chain. However, excessive levels of nutrients from sources such as fertilizer can cause eutrophication, which is the overgrowth of aquatic vegetation. This overgrowth clouds the water and smothers some plants. Over time, excessive nutrient levels can accelerate the natural process by which bodies of water evolve into dry land.

Thermal pollution results from the release of heated water into lakes and streams. Most thermal pollution is generated by power plant cooling systems. Power plants use water to cool their reactors and turbines, and discharge it into lakes and tributaries after it has become heated. Higher water temperatures accelerate biological and chemical processes in rivers and streams, reducing the water's ability to retain dissolved oxygen. This can hasten the growth of algae and disrupt the reproduction of fish. Toxic chemicals and other hazardous materials present the most imminent threat to water quality. The Environmental Protection Agency (EPA) has identified 403 highly toxic chemicals, which are produced by 577 U.S. companies, manufactured in twelve thousand plants, and stored in four-hundred thousand locations across the country. Some chemical plants incinerate toxic wastes, which produces dangerous by-products like furans and chlorinated dioxins, two of the most deadly carcinogens known to the human race. Other hazardous materials are produced or stored by households (motor oil, antifreeze, paints, and pesticides), dry cleaners (chlorinated solvents), farms (insecticides, fungicides, rodenticides, and herbicides), and gas stations and airports (fuel).

The contamination of ground water, rivers, lakes, wetlands, estuaries, and oceans can threaten the health of humans and aquatic life. Sources of water pollution are generally divided into two categories. The first is point-source pollution, in which contaminants are discharged from a discrete location. Sewage outfalls and oil spills are examples of point-source pollution. The second category is non-point-source or diffuse pollution, referring to all of the other discharges that deliver contaminants to water bodies. Acid rain and unconfined runoff from agricultural or urban areas are examples of non-point-source pollution. The principal contaminants of water include toxic chemicals, nutrients and biodegradable organics, and bacterial and viral pathogens.

Water pollution can threaten human health when pollutants enter the body via skin exposure or through the direct consumption of contaminated food or drinking water. Priority pollutants, including dichlorodiphenyl trichloroethane (DDT) and polychlorinated biphenyls (PCBs), persist in the natural environment and bioaccumulate in the tissues of aquatic organisms. These persistent organic pollutants are transferred up the food chain (in a process called biomagnification), and they can reach levels of concern in fish species that are eaten by humans. Finally, bacteria and viral pathogens can pose a public health risk for those who drink contaminated water or eat raw shellfish from polluted water bodies.

Contaminants have a significant impact on aquatic ecosystems. for example, enrichment of water bodies with nutrients (principally nitrogen and phosphorus) can result in the growth of algae and other aquatic plants that shade or clog streams. If wastewater containing biodegradable organic matter is discharged into a stream with inadequate dissolved oxygen, the water downstream of the point of discharge will become anaerobic and will be turbid and dark. Settleable solids, if present, will be deposited on the streambed, and anaerobic decomposition will occur. Over the reach of stream where the dissolved-oxygen concentration is zero, a zone of putrefaction will occur with the production of hydrogen sulfide, ammonia, and other odorous gases. Because many fish species require a minimum of 4–5 mg of dissolved oxygen per liter of water, they will be unable to survive in this portion of the stream.

Direct exposures to toxic chemicals is also a health concern for individual aquatic plants and animals. Chemicals (e.g., pesticides) are frequently transported to lakes and rivers via runoff, and they can have unintended and harmful effects on aquatic life. Toxic chemicals have been shown to reduce the growth, survival, reproductive output, and disease resistance of exposed organisms. These effects can have important consequences for the viability of aquatic populations and communities.

Wastewater discharges are most commonly controlled through effluent standards and discharge permits. Under this system, discharge permits are issued with limits on the quantity and quality of effluents. Water-quality standards are sets of qualitative and quantitative criteria designed to maintain or enhance the quality of receiving waters. Receiving waters are divided into several classes depending on their uses, existing or intended, with different sets of criteria designed to protect uses such as drinking water supply, bathing, boating, fresh-water and shellfish harvesting, and outdoor sports for seawater. For toxic compounds, chemical-specific or whole-effluent toxicity studies are used to develop standards and criteria. In the chemical-specific approach, individual criteria are used for each toxic chemical detected in the wastewater. Criteria can be developed to protect aquatic life against acute and chronic effects and to safeguard humans against deleterious health effects, including cancer. In the whole-effluent approach, toxicity or bioassay tests are used to determine the concentration at which the wastewater induces acute or chronic toxicity effects.

❐❐❐

2

TYPES OF WATER POLLUTION

Introduction

The major sources of water pollution can be classified as municipal, industrial, and agricultural. Municipal water pollution consists of waste water from homes and commercial establishments. For many years, the main goal of treating municipal wastewater was simply to reduce its content of suspended solids, oxygen-demanding materials, dissolved inorganic compounds, and harmful bacteria. In recent years, however, more stress has been placed on improving means of disposal of the solid residues from the municipal treatment processes. The basic methods of treating municipal wastewater fall into three stages: primary treatment, including grit removal, screening, grinding, and sedimentation; secondary treatment, which entails oxidation of dissolved organic matter by means of using biologically active sludge, which is then filtered off; and tertiary treatment, in which advanced biological methods of nitrogen removal and chemical and physical methods such as granular filtration and activated carbon absorption are employed. The handling and disposal of solid residues can account for 25 to 50 percent of the capital and operational costs of a treatment plant. The characteristics of industrial waste waters can differ considerably both within and among industries. The impact of industrial discharges depends not only on their collective characteristics, such as biochemical oxygen demand and the amount of suspended solids, but also on their content of specific inorganic and organic substances. Three options are available in controlling industrial wastewater. Control can take place at the point of generation in the plant; wastewater can be pretreated for discharge to municipal treatment sources; or wastewater can be treated completely at the plant and either reused or discharged directly into receiving waters.

Agriculture, including commercial livestock and poultry farming, is the source of many organic and inorganic pollutants in surface waters and groundwater. These contaminants include both sediment from erosion cropland and compounds of phosphorus and nitrogen that partly originate in animal wastes and commercial fertilizers. Animal wastes are high in oxygen demanding material, nitrogen and phosphorus, and they often harbor pathogenic organisms. Wastes from commercial feeders are contained and disposed of on land; their main threat to natural waters, therefore, is from runoff and leaching. Control may involve settling basins for liquids, limited biological treatment in aerobic or anaerobic lagoons, and a variety of other methods.

Ground Water

Ninety-five percent of all fresh water on earth is ground water. Ground water is found in natural rock formations. These formations, called aquifers, are a vital natural resource with many uses. Rationally, 53% of the population relies on ground water as a source of drinking water. In rural areas this figure is even higher. Eighty one percent of community water is dependent on ground water.

The biggest reason for developing a worldwide effort to monitor and restrict global pollution is the fact that most forms of pollution do not respect national boundaries. The first major international conference on environmental issues was held in Stockholm, Sweden, in 1972 and was sponsored by the United Nations (UN). This meeting, at which the United States took a leading role, was controversial because many developing countries were fearful that a focus on environmental protection was a means for the developed world to keep the undeveloped world in an economically subservient position.

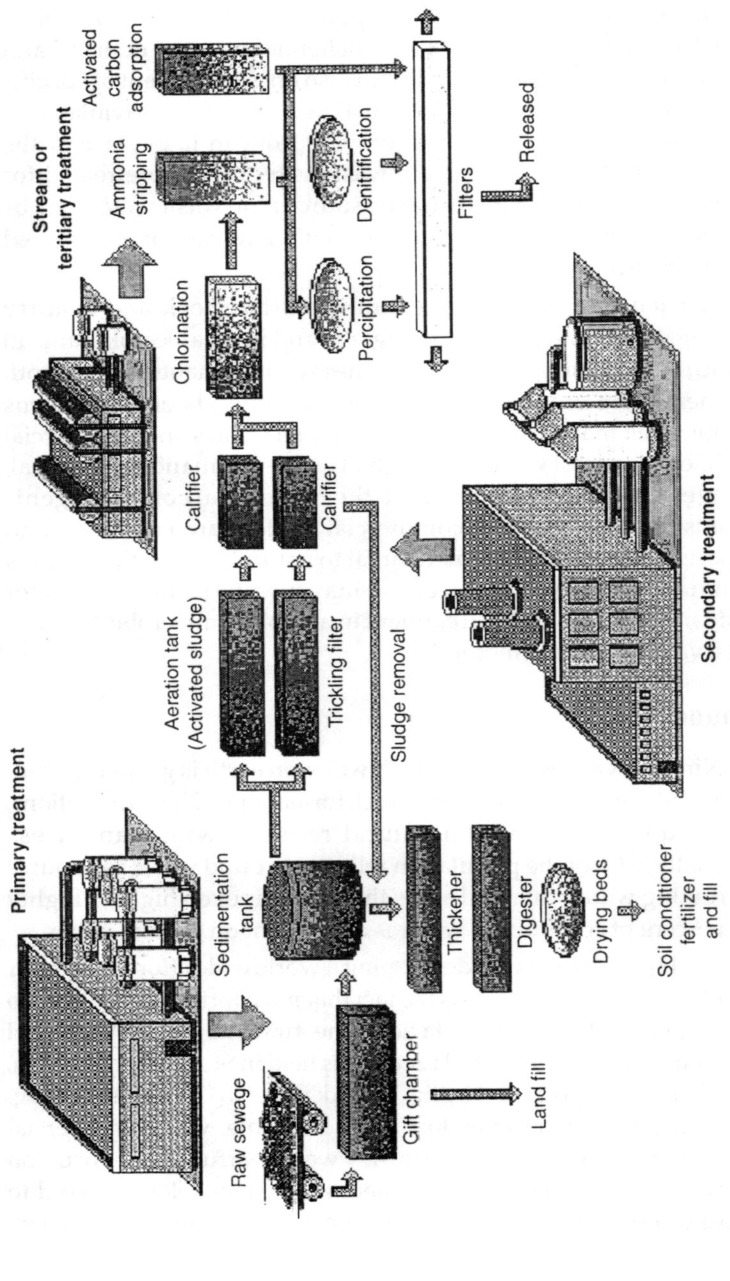
Primary treatment
Raw sewage
Gift chamber
Land fill
Sedimentation tank
Thickener
Digester
Drying beds
Soil conditioner fertilizer and fill
Aeration tank (Activated sludge)
Trickling filter
Sludge removal
Calrifier
Calrifier
Secondary treatment
Chlorination
Stream or teritiary treatment
Ammonia stripping
Activated carbon adsorption
Percipitation
Denitrification
Filters
Released

UNEP was designed to be:

- the environmental conscience of the United Nations,
- and, in an attempt to allay fears of the developing world, it became the first UN agency to be headquartered in a developing country, with offices in Nairobi, Kenya. In addition to attempting to achieve scientific consensus about major environmental issues, a major focus for UNEP has been the study of ways to encourage sustainable development increasing standards of living without destroying the environment. At the time of UNEP's creation in 1972, only 11 countries had environmental agencies. Ten years later that number had grown to 106, of which 70 were in developing countries.

Distribution of World's Water

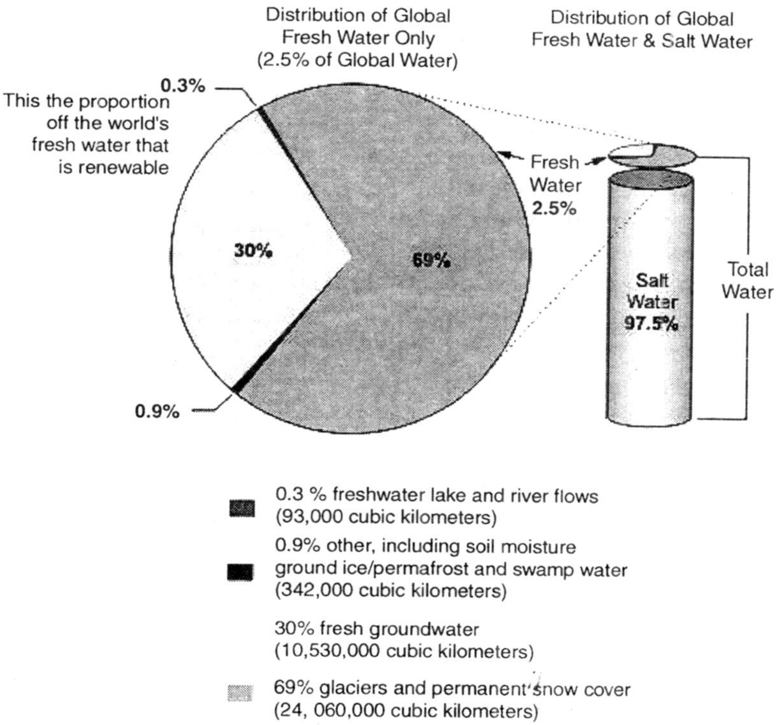

Water Quality

Water quality is closely linked to water use and to the state of economic development. In industrialized countries, bacterial contamination of surface water caused serious health problems in major cities throughout the mid 1800s. By the turn of the century, cities in Europe and North America began building sewer networks to route domestic wastes downstream of water intakes. Development of these sewage networks and waste treatment facilities in urban areas has expanded tremendously in the past two decades. However, the rapid growth of the urban population (especially in Latin America and Asia) has outpaced the ability of governments to expand sewage and water infrastructure. While waterborne diseases have been eliminated in the developed world, outbreaks of cholera and other similar diseases still occur with alarming frequency in the developing countries. Since World War II and the birth of the ?chemical age?, water quality has been heavily impacted worldwide by industrial and agricultural chemicals. Eutrophication of surface waters from human and agricultural wastes and nitrification of groundwater from agricultural practices has greatly affected large parts of the world. Acidification of surface waters by air pollution is a recent phenomenon and threatens aquatic life in many area of the world. In developed countries, these general types of pollution have occurred sequentially with the result that most developed countries have successfully dealt with major surface water pollution. In contrast, however, newly industrialized countries such as China, India, Thailand, Brazil, and Mexico are now facing all these issues simultaneously.

Water Cycle and Pollution

Water cycle is, simply saying, the circulation of water in earth. In fact, the water in the earth's biosphere is used and reused again and again. This is called water cycle or continuous movement of water between the earth and the atmosphere. It involves the following mechanisms:

- Evaporation: changing of water from liquid to gas
- Transpiration: Release of water vapor from plant leaves
- Condensation: changing of vapor to liquid (cooled down)
- Precipitation: Water that returns to the earth (water droplets in clouds become large enough and there comes the rain).

What's the relation of water cycle and pollution?

According to the water cycle, naturally, water around us will be absorbed to the land (soil) and rivers will stream from the upstream to the downstream and released to the sea. In normal situation organic pollutants are biodegraded by microbes and converted to a form that brings benefits to the aquatic life. And for the inorganic pollutants, in the same situation, don't bring to much hazards because they are widely dispersed and have almost no effect to the environment which they are released to.

In a small scale, both inorganic and organic pollutants safely decompose throughout the stream, their concentration decrease in the sea, and they don't harm the sea ecosystem and its distribution. But in an excessive scale, communities in beach and estuary will be affected by the pollutants, and can heavily harm them.

Other Sources of Water Pollution

Towns and municipalities are also major sources of water pollution. In many public water systems, pollution exceeds safe levels. One reason for this is that much groundwater has been contaminated by wastes pumped underground for disposal or by seepage from surface water. When contamination reaches underground water tables, it is difficult to correct and spreads over wide areas. In addition, many U.S. communities discharge untreated or only partially treated sewage into the waterways, threatening the health of their own and neighboring populations.

Along with domestic wastes, sewage carries industrial contaminants and a growing tonnage of paper and plastic refuse. Although thorough sewage treatment would destroy most disease-causing bacteria, the problem of the spread of viruses and viral illness remains. Additionally, most sewage treatment does not remove phosphorus compounds, contributed principally by detergents, which cause eutrophication of lakes and ponds. Excreted drugs and household chemicals also are not removed by present municipal treatment facilites, and can be recycled into the drinking water supply.

Rain drainage is another major polluting agent because it carries such substances as highway debris (including oil and chemicals from automobile exhausts), sediments from highway and building construction, and acids and radioactive wastes from mining operations into freshwater systems as well as into the ocean. Also transported by rain runoff and by irrigation return-flow are animal wastes from farms and feedlots, a widespread source of pollutants impairing rivers and streams, groundwater, and even some coastal waters. Antibiotics, hormones, and other chemicals used to raise livestock are components of such animal wastes. Pesticide and fertilizer residues from farms also contribute to water pollution via rain drainage.

Ocean Pollution

Large and small craft significantly pollute both inland and coastal waters by dumping their untreated sewage. Oil spilled accidentally or flushed from tankers and offshore rigs (900,000 metric tons annually) sullies beaches and smothers bird, fish, and plant life. In 1989 in one of the world's worst single instances of water pollution, the *Exxon Valdez* spilled 11 million gallons of oil in Prince William Sound, Alaska, causing great environmental destruction. In 1997, the 22 oil spills reported worldwide involved a total of 15 million gallons (57 million liters) of oil. In addition to its direct damage to wildlife, oil takes up fat-soluble poisons like DDT, allowing them to be concentrated in organisms that ingest the oil-contaminated water; thus such poisons enter the food chains leading to sea mammals and people.

Both DDT, which has been banned in the United States since 1972, and PCBs are manufactured in many parts of the world and are now widespread in the Atlantic and Pacific oceans. In addition, tarry oil residues are encountered throughout the Atlantic, as are styrofoam and other plastic rubbish. Plastic bits litter sections of the Pacific as far north as Amchitka Island near Alaska. Garbage, solid industrial wastes, and sludge formed in sewage treatment, all commonly dumped into oceans, are other marine pollutants found worldwide, especially along coastal areas.

Dangers of Water Pollution

Virtually all water pollutants are hazardous to humans as well as lesser species; sodium is implicated in cardiovascular disease, nitrates in blood disorders. Mercury and lead can cause nervous disorders. Some contaminants are carcinogens. DDT is toxic to humans and can alter chromosomes. PCBs cause liver and nerve damage, skin eruptions, vomiting, fever, diarrhea, and fetal abnormalities. Along many shores, shellfish can no longer be taken because of contamination by DDT, sewage, or industrial wastes.

Dysentery , salmonellosis , cryptosporidium , and hepatitis are among the maladies transmitted by sewage in drinking and bathing water. In the United States, beaches along both coasts, riverbanks, and lake shores have been ruined for bathers by industrial wastes, municipal sewage, and medical waste. Water pollution is an even greater problem in the Third World, where millions of people obtain water for drinking and sanitation from unprotected streams and ponds that are contaminated with human waste. This type of contamination has been estimated to cause more than 3 million deaths annually from diarrhea in Third World countries, most of them children.

Wastewater Treatment

Raw sewage includes waste from sinks, toilets, and industrial processes. Treatment of the sewage is required before it can be safely buried, used, or released back into local water systems. In a treatment plant, the waste is passed through a series of screens, chambers, and chemical processes to reduce its bulk and toxicity. The three general phases of treatment are primary, secondary, and tertiary. During primary treatment, a large percentage of the suspended solids and inorganic material is removed from the sewage. The focus of secondary treatment is reducing organic material by accelerating natural biological processes. Tertiary treatment is necessary when the water will be reused; 99 percent of solids are removed and various chemical processes are used to ensure the water is as free from impurity as possible.

Conclusion

It is concluded that approximately 40% of the nation's surveyed lakes, rivers, and estuaries were too polluted for such basic uses as drinking supply, fishing, and swimming. The pollutants include grit, asbestos , phosphates and nitrates , mercury , lead , caustic soda and other sodium compounds, sulfur and sulfuric acid , oils, and petrochemicals .

In addition, numerous manufacturing plants pour off undiluted corrosives, poisons, and other noxious byproducts. The construction industry discharges slurries of gypsum, cement, abrasives, metals, and poisonous solvents. Another pervasive group of contaminants entering food chains is the polychlorinated biphenyl (PCB) compounds, components of lubricants, plastic wrappers, and adhesives. In yet another instance of pollution, hot water discharged by factories and power plants causes so-called thermal pollution by increasing water temperatures. Such increases change the level of oxygen dissolved in a body of water, thereby disrupting the water's ecological balance, killing off some plant and animal species while encouraging the overgrowth of others.

3

Nature of Waste Discharges and their Effects

A complete description of the physical, chemical, and biological aspects of water quality, as a function of natural and man-made factors, and the technological measures available for changing water quality can easily fill several extensive volumes. The aim is to provide a background for consideration of the economic and institutional aspects of water quality management.

Each type of receiving water has characteristics that, to some degree, result in differential impacts of the same waste discharge on water quality, no attempt at complete coverage is made. It describes the relationships between source water quality (quality in the watercourse) and various water uses, including intake water treatment methods and costs. It considers various methods for improving water quality, subsumed under the two major categories of reducing wastes prior to discharge and improving or making more effective use of the assimilative capacity of watercourses. The relation of these various methods of region-wide water quality and water resources management will also be indicated.

Major Types of Wastes and their Sources

The many substances having an impact on water quality that enter surface watercourses and groundwater aquifers as a result of man's domestic, industrial, agricultural, and recreational activities can be grouped in various ways. One common classification differentiates between physical substances (e.g., radioactive, inorganic, organic) and biological substances (e.g., bacteria and viruses). In addition, the quality of a waste discharge can be measured in various ways. Some of the common measures are biochemical oxygen demand (BOD), chemical oxygen demand (COD), alkalinity, acidity, color, hardness, taste, odor, turbidity, and electrical conductance.

For the purposes of this book, however, the most useful classification of wastes distinguishes between nondegradable and degradable substances, based on their behavior in receiving waters. Nondegradable wastes are usually diluted and may be changed in form, but they are not appreciably reduced in weight in the receiving water. Degradable wastes are reduced in weight by the biological, physical, and chemical processes which occur in natural waters.

Nondegradable Wastes

The sources of nondegradable substances mainly inorganic chemicals, such as chlorides, synthetic organic chemicals, and inorganic suspended solids are many and widespread. Industrial waste discharges frequently contain inorganic or metallic salts, synthetic organic chemicals, and other materials which may be toxic or corrosive, may impart color or taste to the water, and or may produce odors. Domestic water use results in a small increase in the content of chlorides and other dissolved salts. The return flow from irrigation is generally significantly higher in dissolved salts than the applied irrigation water. The discharge of copper, zinc, uranium, and other compounds in drainage from mine tailings can have very adverse impacts on water quality, and acid mine drainage is an acute problem in some of the major coal-producing areas. In the heavily industrialized sections of the country and in the major irrigated areas of the West, industrial operations and irrigated agriculture may be major sources of chlorides and other dissolved salts.

Natural geologic formations, such as salt deposits, may result in high concentrations of chlorides and or dissolved solids in watercourses. Another instance of deterioration of water quality by a natural source is from seawater intrusion in groundwater aquifers near coastal areas (following a change in the hydraulic gradient because of pumping from the aquifers). Similarly, the quality of water in inland groundwater aquifers can be adversely affected by excessive pumping where there are hydraulic connections with brackish or saline aquifers.

Suspended materials, which can be considered nondegradable in the sense that the), are essentially not changed in weight in receiving waters, consist of sediment from natural and accelerated erosion from land surfaces and stream channels and colloidal materialfrom clay deposits and in domestic and industrial waste discharges. Suspended sediment usually settles out, but colloidal matter does not, except under unusual circumstances. Both suspended sediment and colloidal matter cause turbidity in surface waters. This not only makes the water less attractive, but can damage fish life. Turbidity may also inhibit the growth of algae and other aquatic plants, which may or my not be undesirable.

Agricultural activities generate both nondegradable and degradable wastes. Storm runoff from agricultural land carries silt, clay, fertilizers, and weedicides, as well as organic material and bacteria, into watercourses. In contrast to the sources of wastes described above, which result in waste discharges concentrated at one or a few points, agricultural activities generally result in dispersed waste loads, except in the case of feed lot operations. In predominantly agricultural regions, for example the Missouri Basin, these dispersed waste loads may be the major sources of nitrogen, phosphate, and pesticides, as well as organic wastes.

Actually the classification into degradable and nondegradable waste loads is a simplification. Some types of substances, such as radioactive materials and some of the organic chemicals, are degradable to some degree. Viruses also appear to be in the in-between category. Some radioactive wastes decay rapidly; others extremely slowly. Some of the synthetic organic compounds, such as alkyl benzene sulfonate (ABS) detergents, are not strictly nondegradable, but they resist attack by stream biota. On the other hand, the new detergents based on linear alkylate sulfonate (LAS) are degradable, both in streams and other watercourses and in waste treatment plants.

One other point with respect to nondegradable wastes merits mention, namely, the problem of identifying the substances in the receiving water. With respect to many of the complex industrial, agricultural, and pharmaceutical chemical compounds, not only have techniques not yet been developed to identify them, but the concentrations involved are often on the order of a few parts per billion, rather than parts per million as with chlorides or suspended sediment. Further, the extent to which there are synergistic or antagonistic reactions is not fully understood as yet. Viruses also present identification problems because they can exist in a dormant condition outside living organisms and be reactivated under proper environmental conditions.

Degradable Wastes

Degradable wastes the substances that are reduced in quantity by the biological, chemical, and physical phenomena characteristic of natural waters include organic wastes of various kinds from domestic and industrial operations, bacteria, and thermal discharges. The most widespread source of degradable organic materials is domestic sewage. In terms of total quantity, industrial discharges represent the largest source. The highly unstable, putrescible, organic waste can be converted to stable inorganic materials (bicarbonates, nitrates, sulphates, and phosphates) by the bacteria and other organisms found in natural water bodies. If the water is not too heavily loaded, the process, commonly but somewhat deceptively known as self-purification, will proceed aerobically by the action of bacteria utilizing free oxygen and will not produce offensive odors. If the receiving waters are loaded excessively, the process of degradation will proceed anaerobically by the action of bacteria not utilizing free oxygen, and hydrogen sulfide and other gases will be produced. Both of these processes are utilized in waste treatment plants, which, in essence, systematize, control, and accelerate the processes that take place in natural waters.

Bacteria are omnipresent in natural waters. Except for a few types, i.e., the pathogens, the vast majority are beneficial. The primary waste discharges that are sources of bacteria are from activities of warmblooded animals. The coliform count is the most common index of such waste discharges. Total bacteria count is also used as an indicator of water quality in relation to such discharges. Both coliform count and total bacteria count serve as crude proxies for the actual substances of concern, namely, the bacteria that cause infectious diseases primarily typhoid, dysentery, and cholera. Bacteria may be considered degradable since they tend to die off rather quickly after leaving the body.

Thermal loads are classed as degradable because the heat is dissipated in receiving waters, primarily by evaporation in surface water, and by some combination of evaporation and conduction in groundwater. The primary sources of thermal loads are the generation of electrical energy and the cooling operations in the petroleum refining, pulp and paper, iron and steel, and chemical manufacturing industries. Thermal discharges are particularly troublesome when industrial operations are concentrated along a stream or an estuary.

Effects of Waste Discharges on Receiving Waters

The weight of a waste in a discharge and the volume of the discharge, the concentration of a nondegradable waste over time in a stream or lake is relatively easy to predict because dilution is the dominant process when the substance is in solution. Prediction is more difficult for estuaries because of the complex hydraulic characteristics resulting from tidal action and topography. Inorganic chemicals may precipitate out in lakes, reservoirs, and, under certain conditions, in flowing streams. Suspended sediment will settle out to some degree in receiving waters, particularly in sections where there is a major decrease in stream velocity, although the spatial pattern of deposition cannot always be accurately predicted. For a nondegradable waste discharged to a groundwater aquifer, predicting the concentration over time at various distances from the original discharge point is complicated by variations in transmissibility, both vertically and horizontally, and by the physical processes (such as adsorption) and chemical reactions which are possible. In all of these situations the weight of the discharged substance remains essentially unchanged, although the form may be modified in the receiving waters.

It should be emphasized that, while the prediction of concentration of a nondegradable waste in a specified water body is relatively easy for a known weight of waste and given volume of receiving water, prediction of the actual time pattern of concentration resulting from a waste discharge is far less simple. This is true for two reasons. First, with respect to the receiving water, there is uncertainty about the quantity of water available for dilution, i.e., hydrologic uncertainty. Hydrologic events are stochastic variables, hence the quantity of water available is known only in terms of a probability distribution. Second, where the source of the nondegradable substance is an industrial operation, there often is significant variation from day to day (and even within the day) in the quantity of waste discharged. Despite these difficulties, procedures have been developed which enable reasonably accurate prediction of time patterns of concentrations resulting from defined time patterns of discharge of nondegradable wastes.

Degradable Wastes

Predicting the time pattern and spatial pattern of concentration of wastes presents considerably more difficult technical problems when the wastes are degradable than when they are nondegradable. This is particularly true for organic wastes, but it holds to a lesser degree for thermal discharges, for certain types of radioactive wastes, and for those synthetic organic pesticides which decompose to some degree. With respect to organic and thermal waste discharges, the primary concern is not the fate of the wastes *per se*, but the effect of their degradation on water quality, in particular the dissolved oxygen concentration in the receiving waters. Because of the importance of organic waste discharges in relation to total waste discharges and because of their widespread occurrence, we turn to a relatively detailed discussion of such wastes.

A measure of organic waste load is biochemical oxygen demand (BOD), which indicates the amount of oxygen drawn upon in the process of decomposition of the waste. The amount of oxygen demanded and the rate at which it is drawn upon are functions of the type and quantity of the waste and of other factors, among which the most important are the chemical characteristics and the temperature of the receiving water. Toxic substances, for example, may appreciably reduce the rate of decomposition by inhibiting bacterial action, and in extreme cases of toxic concentration a body of water may become bacteriologically "dead." At higher temperatures, when the oxygen saturation level of water is relatively low, bacterial action is accelerated, wastes are degraded more rapidly, and dissolved oxygen in the water is drawn upon more rapidly. The imbalance between available oxygen and oxygen demand may proceed to the point where septic (anaerobic) conditions result. Such conditions are most likely to occur in summer and early fall when streamflows tend to be low, and temperatures high.

The rate at which BOD is exerted combined with the rate at which oxygen is restored determines the level of dissolved oxygen (D.O.). In flowing water the combined effect of an organic waste discharged at a specific location and reaeration in the stream results first in a decrease and then in an increase in D.O. as the waste is carried or moved downstream. This phenomenon is illustrated by a characteristic curve known as the "oxygen sag." The variables affecting the shape of the curve can be described by standard equations.

Other things being equal, factors that reduce the rate of BOD lengthen and flatten the oxygen sag, while those that accelerate BOD have the reverse effect. The shape of the oxygen sag is also affected by the rate of reaeration, which depends largely on channel characteristics as they affect turbulence and the area of the air-water interface, the net photosynthetic oxygen production, and the velocity of streamflow.

The role of photosynthesis, as a producer of oxygen, in the dissolved oxygen balance of receiving waters is a complex one. Photosynthesis depends on light and nutrients available, stream velocity, amounts and types of waste loads, water depth, and water temperature. Most of the photosynthesis in water bodies is performed by aquatic green plants called algae, which range in size from tiny, single-cell plants to various types up to several feet in length.

The contribution of photosynthesis to the oxygen content is complicated by the problem of respiration. An important factor in the assessment of this contribution is the P/R ratio, the ratio between oxygen production and community respiration, where community includes aquatic animals, plants, and bacteria. Directly downstream from the discharge of an organic load the P/R ratio is usually very small, i.e., on the order of 0.01. The peak production of oxygen occurs in the early portion of the "zone of recovery", where the P/R ratio may be on the order of 3.0 or more. Downstream a significant distance from the waste discharge, as the dissolved oxygen content of the stream approaches the level prior to the waste discharge, the P/R ratio tends to approach unity. The wide variation in the P/R ratio which has been found in various streams complicates the analysis of the impact of organic waste loads on dissolved oxygen.

A further complicating factor with respect to organic waste loads and photosynthesis involves the diurnal variation in dissolved oxygen. The diurnal variation may be of particular importance for certain types of water uses, such as fisheries. Thus the analysis of water quality based solely on longer time periods, i.e., weeks or months, may in certain instances—be inadequate as a basis for optimal water quality management.

The low point on the oxygen sag (e.g., D_c) has been a governing factor in planning or determining the size or capacity and the design of waste treatment plants. Of course, "point" in this context actually means a particular reach or an area of a given stream, estuary, or lake. Sanitary engineering practice has typically been to calculate the oxygen sag at a given "low" level of streamflow and to determine the capacity and design of a waste treatment plant which would result in a specified level of dissolved oxygen in the critical reach of the oxygen sag. The stochastic nature of streamflow and its effect on the variation in assimilative capacity of the stream, and hence on the possibilities for more flexible treatment facilities, have generally been ignored in this planning process, at least until recently.

The equations detailed in the appendix to this chapter were first devised to describe the behavior of D.O. in a single reach of stream subject to a single waste discharge. They can, however, be applied to multiple reaches, since the deficit at the downstream end of one reach is also the initial deficit (D_a) at the upstream end of the next reach. Such straightforward use of the equations is possible only under highly simplified conditions, including a constant oxygen saturation along the length of the stream. The basic equations describing D.O. behavior have been elaborated in a number of ways to handle more complex situations including both multiple reaches and multiple points of waste discharge.

Consideration of waste disposal and water quality management on a regional basis requires methods for tracing the effect of waste discharges on long reaches of streams, in lakes, and throughout estuaries and groundwater basins. Where wastes are discharged to flowing streams from closely spaced outfalls, the level of D.O. at any given point is a function of the rate of reaeration and the exertion of BOD from the various sources, and of course the factors that affect both. (The same is true for lakes and estuaries.) At a given downstream point the BOD from the farthest upstream sources may be almost completely exhausted, while that from less distant sources will still be exerted at a high rate. A dissolved oxygen profile for a long stretch of such a stream will show deep valleys below large waste discharges or closely spaced small discharges and peaks at or near the D.O. saturation level at other locations. It shows a dissolved oxygen profile of the Ohio River obtained by computer simulation for given conditions of flow, temperature, and waste loads.

Among the complexities that beset the calculation of oxygen levels, especially for long reaches of a stream, is the fact that BOD proceeds in two stages. When an organic waste is discharged into a stream where dissolved oxygen is near saturation level, there is an immediate and major draft upon the D.O. content of the stream, as the organic wastes are degraded by bacterial action. Thereafter the D.O. level tends to recover. Farther downstream, corresponding roughly to a travel time of five to seven days, a "second stage" BOD occurs, as the nitrogen embodied in the organic wastes is converted to nitrite and then to nitrate by aerobic nitrifying bacteria. The second stage is more diffuse and does not by itself reduce the D.O. to as low a level as the first stage does.

Tracing the fate of an organic waste load in an estuary is even more difficult because the estuarine environment is complicated by tidal action, the large expanse of air-water interface, more complex hydraulic characteristics, and generally complicated patterns of inflows. In addition, photosynthesis may play an even larger role in the oxygen balance of an estuary than of a flowing stream.

Despite these complexities, significant advances have been made in developing models which interrelate the many variables involved, such as streamflow and temperature with less or more consideration of the stochastic nature of both and multiple waste loads, and enable prediction of regional profiles of D.O., temperature, and other water quality characteristics. These analytical models, virtually all of which depend on the use of computers, have provided the bases for economic optimization studies of water quality management systems.

To cite one example, Thomann developed a mathematical representation of the Delaware estuary water quality system, composed of two subsystems: (1) the biochemical oxygen demand subsystem; and (2) the dissolved oxygen subsystem. For each of the sections into which the estuary was divided, mass balance equations were written for both subsystems, assuming steady-state systems, i.e., invariant with time. These consisted of two linear differential equations based on the hydrological, physical, and biochemical characteristics in each section. The result was two sets of thirty simultaneous equations.

By matrix manipulation a set of transfer functions was obtained from the coefficients of these equations. This set of transfer relationships yielded the transformation of a waste load input in any specific section to the dissolved oxygen level in any other section. Since all of the equations in the system are linear, the solutions could be linearly superimposed. Thus, the total effect at any section was found by adding the effects in that section caused by inputs at any number of sections in the estuary.

Then the assumptions of a stationary system were removed, i.e., the parameters in the equations were permitted to vary with time, and a digital computer was used to solve the equations. Comparisons between the model results and actual data provided verification of the model and simultaneously the more accurate specification of the variables in the equations. The verified model was the basis for analyses of various combinations of inflow to the estuary, waste discharges, and additions of supplemental oxygen, and variations in reaeration and degradation rates.

The basic phenomena involved in the dissolved oxygen-organic waste load system of the Delaware estuary were reaeration, degradation, and tidal diffusion. By modifying the model, e.g., eliminating the degradation process, the behavior of nondegradable wastes, such as chlorides, could be investigated. The degradation of organic wastes discharged to groundwater occurs primarily by anaerobic decomposition. Detailed studies of the dispersion and degradation of organic wastes discharged to porous media have been made.

Lakes are another special case. Being more stationary than streams or estuaries, they are naturally more susceptible to eutrophication (nutrient buildup). This process is speeded up by the discharge of organic wastes, even residual wastes after significant levels of treatment. Nutrient enrichment tends to be accompanied by oxygen deficiencies and can have a serious effect on water quality, as evidenced by conditions in Lake Tahoe (California-Nevada), and Lake Erie. Eutrophication, the complex circulation patterns of large lakes, stratification and temperature inversions, all combine to make the analysis of the effect of organic waste discharges on water quality in lakes very difficult. The attention presently being devoted to studies of lacustrine environments suggests that progress should be forthcoming in developing methods for simulating the behavior of the many interrelated variables in such environments.

The residual products of organic waste degradation in water bodies are plant nutrients, i.e., nitrogen, phosphorus, carbon. Under propitious conditions of water temperature and water clarity, these plant nutrients contribute to algae growth, which in turn affects the oxygen balance in the receiving water. If algae are present in large quantities, they may affect the appearance, taste, and odor of the water.

As can be deduced from the previous discussion, degradable wastes in the form of thermal discharges are of importance primarily because of their effect on water temperature, which in turn affects both the rate of reaeration and the rate of decomposition of organic wastes, and hence affects D.O. As in the case of organic waste discharges, methods have been developed to predict the dissipation of thermal discharges in water bodies.

Conclusion

The simple classification systems for types of wastes and types of receiving waters comprise a convenient point of departure for discussing the technical background for water quality management, they obscure, to some extent, the highly complex phenomena and interrelationships which are involved. A vast number of substances enter watercourses. Some result from man's activities and others are of natural origin. Their effects are interrelated in many complicated ways. For example, water clarity is an important precondition for the production of algae. Thus, streams with high turbidity are not likely to have the taste and odor problems that often accompany prolific algae growth. Consequently, when measures are taken to reduce turbidity, thereby increasing the clarity of water, algae production may increase with resulting taste and odor problems that previously did not exist.

Lakes, reservoirs, and estuaries, like flowing streams, are subject to the deposition of sludge banks, which if suddenly dispersed produce a "shock load" of oxygen demand. But these bodies of water also have special problems of their own. Stratification, the formation in essentially stationary bodies of water of thermal layers which prevent deep circulation, complicates the analysis of water quality in lakes and reservoirs. Plant nutrients also take on special significance because they tend to accumulate over time in relatively stationary bodies of water and contribute to the depletion of oxygen in the lower levels. In a tidal estuary, wastes may be moved back and forth within the estuary for long periods of time rather than being dispersed into the ocean. Thus, residual wastes may build up with resulting water quality deterioration.

Much has been done to develop generalizations about the effect of waste discharges on the quality of various types of receiving water. However, the accuracy and precision are less and the uncertainty is greater in relation to predicted behavior of these phenomena than in many fields of scientific and engineering analysis. Therefore, it is important that any estimates of behavior based upon simplified models be checked with empirical data. Nevertheless, much progress has been made in developing mathematical models of waste transport, dilution, and degradation processes in receiving waters. Several models are now operational and capable of predicting, with a reasonable degree of accuracy, the sequence of results over time and space of the discharge of wastes. Such models are essential for economic optimization analyses of water quality management systems, and as tools for water quality management organizations.

❑❑❑

4

Managing Waste Loads and their Assimilative Capacity

Introduction

The quality of water bodies can be improved by reducing the waste loads discharged to them, by increasing or making more effective use of their assimilative capacity, and by combinations of the two. If the focus is on water quality management in streams, and if there is no problem with floating materials, reducing waste loads and changing assimilative capacity are technical substitutes for one another. Assimilative capacity can be increased by altering the time patterns of streamflows, by altering the time pattern and/ or locations of waste discharges, by artificial reaeration, or by some combination of these measures. If the assimilative capacity is known, intake water treatment can be a partial substitute for waste load reduction. Indeed, if water were not useful, offensive, or altered in quality while *in situ*, intake water treatment would be technically interchangeable with waste load reduction. Flowing streams, of course, are much more amenable to procedures for changing assimilative capacity than are groundwater aquifers or even reservoirs and lakes.

Improving the water quality of receiving waters by reducing waste loads can be accomplished in two broad ways: first, by reducing the generation of wastes; and second, by modifying the residual wastes. Within these two broad categories a variety of techniques are available, as indicated in Table 4.1. The dividing lines between methods that reduce waste generation and those that affect the residual wastes are not always as clearcut as implied in the table. Materials recovery, for example, is in a sense a change in production process because it involves a change in the mix of raw product inputs; process change may be necessary before in-plant water recirculation is possible; and product output changes may require some changes in production processes. Nevertheless, the distinctions are worth preserving as a basis for discussion. Also listed in Table 4.1 are ways of managing the assimilative capacity of receiving waters. For example, following generation, waste materials may be recovered, used for the production of byproducts, reduced by waste treatment, temporarily stored prior to discharge, conveyed to another water user, or discharged in various possible ways into one or more receiving water bodies.

Table 4.1. Methods for Improving the Quality of Receiving Waters

METHODS FOR REDUCING WASTE DISCHARGES

Methods for reducing wastes generation

1. Change in type of raw material inputs
2. Change in production process
3. Change in product outputs
4. In-plant recirculation of water

Methods for reducing wastes after generation

1. Materials recovery
2. By-product production
3. Waste treatment
4. Effluent reuse (including groundwater recharge, wastewater reclamation or renovation)

METHODS FOR INCREASING OR MAKING BETTER USE OF ASSIMILATIVE CAPACITY

1. Addition of dilution water
2. Multiple outlets from reservoirs
3. Reservoir mixing
4. Reaeration of streams
5. Saltwater barriers
6. Effluent redistribution (including regulated discharge)

Industrial wastes play such a large role in water quality management that an understanding of the factors that influence the generation of these wastes is extremely important. This is not to say that the magnitude of wastes generated by domestic and commercial users is insignificant. On the contrary, with expanding population, particularly in urban areas, the absolute quantity of domestic wastes generated is bound to increase.

Flow Diagram of Water Utilization, Waste Generation, and Waste Disposal

I = Intake water.

R = Recirculated water.

G = Gross water applied for all in-plant uses.

U = Consumptive use or net depletion of water.

$U = U_P + U_D + U_R$, where

U_P = consumptive use in the production process,

U_D = consumptive use in the wastewater disposal system, and U_R = consumptive use in the recirculation system.

D = Wastewater discharge from the production process.

E = Final effluent from the production unit (available for reuse). Where a lagoon or spray irrigation system is involved, the final effluent (if any) consists of seepage, lagoon discharge, and/or surface runoff.

W = Waste load generated in all operations of the production process, for example, pounds of BOD, Btu's of heat.

W_E = Waste load in the final effluent ie., pounds of BOD, Btu's

Degree of recirculation = $R/G \times 100\%$.

And with the widespread adoption of garbage and paper grinders, there may also be an increase in the generation of wastes per capita within the home. But industrial waste discharges currently account for the greater share of the residual materials discharged to receiving waters in the United States, and their control is a complex, multi-faceted process. In contrast, treatment is the only general method for reducing waste discharges from households, barring significant changes in consumption patterns. Accordingly, the primary focus is on industrial wastes.

Improving Water Quality by Reducing Waste Discharges

Factors Reducing Wastes Generation

The generation of industrial wastes per unit of raw product processed or per unit of final product is a function primarily of the type of raw materials used in the production process, the technology of the production process, the product mix, and sometimes the extent of in-plant water recirculation. Thus there is a wide range in the wastes generated per unit by different industries, and also by different plants within a single industry. To illustrate, a German study reported ratios of high to low waste discharge (in pounds of BOD per unit of product or raw product processed) within several industries as follows: about 25 for paper mills, 2.5 in malt factories, 6 in starch factories, 50 in pork slaughterhouses, about 10 in tanneries, and about 20 in textile factories.

Shifting to a different grade or type of raw product is one way of reducing waste loads. The sulfur content of crude oil has an important bearing on a petroleum refinery's waste. In pulp and paper, the type of wood used is significant. In fruit and vegetable canning, the size, shape, per cent of solids, and resistance to damage and bruising in harvesting and processing the raw product affect the waste load per ton of raw product processed.

Changes in production processes, however, have the major impact on wastes generation, although it should be emphasized that technological changes have not been, and are not usually, instituted because of water quality problems. In fact, most changes in production technology have been stimulated by factors unrelated to water problems, and have been developed without explicit consideration for their effects on water quality. On the other hand, where various stimuli to management, such as sewer charges, have been used, they have often resulted in process modifications. As more systematic means are developed to bring to bear on industry the broader social costs associated with the discharge of waste materials into the environment, the generation and control of wastes will receive more prominent consideration in process design, and perhaps even in product development. Studies of several industries make clear that even comparatively small design changes can have a substantial effect on the generation of wastes.

In the pulp and paper industry the proportions of the total output produced by different processes have shifted significantly over the past two decades. The largest and most significant shift has been from the sulfite process to the sulfate process, which, by making the recovery of chemicals for internal reuse economic, can reduce the waste load in terms of pounds of BOD per ton of product to about 5 or 10 per cent of the previous level. The development of save-alls to recover waste fibers which formerly were simply discharged into water courses was stimulated by the increase in the value of fiber. Their use has resulted in a significant reduction in terms of pounds of both BOD and suspended solids per ton. Both black liquor recovery systems and save-alls were stimulated originally by the economic incentive to save materials useful in the production process, rather than by waste problems.

Table 4.2. Production of Wood Pulp in the United States, by Process and Year (per cent of total production)

Process	1960	1965	1970	1975	1980	1985
Mechanical	18.2	18.0	14.9	13.2	13.0	11.8
Unbleached sulfite	11.1	8.0	5.0	3.1	2.1	1.5
Bleached sulfite	18.0	15.2	14.2	13.9	12.6	11.4
Unbleached sulfate	35.3	35.6	38.4	37.0	34.3	35.4
Bleached sulfate	6.5	8.4	12.1	17.5	23.3	26.3
Semi-chemical			4.6	6.8	7.9	8.7
Soda	5.9	4.2	3.5	2.1	1.7	0.7
All other	4.9	10.6	7.2	6.4	5.2	4.4
Total	99.9	100.0	99.9	100.0	100.1	100.2

Other possibilities for process changes which will reduce wastes generated appear to be on the horizon. One of these is a "dry" papermaking process that will use considerably less water as a defibrating and distributing agent in the paper-making process.

A particularly difficult waste problem in the steel industry for many years has been disposal of the "waste pickle liquor" generated when steel is pickled with sulfuric acid. Attempts to develop an economically feasible sulfuric acid recovery process have met with little success under U.S. conditions. Another solution is to shift to hydrochloric acid (HCI) in the pickling process itself. A recent report indicates that the acid losses in the effluent are reduced by this process change from about 13 pounds of sulfuric acid per ton of steel pickled to about 0.2 pound of hydrochloric acid per ton of steel. Hydrochloric acid offers other advantages as well a high-quality product (in terms of brightness of surface) and faster production.

In the beet sugar industry the waste load generated in pounds of BOD per ton of beets processed has been reduced greatly in the last two decades by comparatively simple and economical changes in processes. The substitution of drying of beet pulp for storage of wet beet pulp in silos and the use of Steffens waste for the production of by-products, or the complete elimination of Steffens processing can reduce BOD generation by about 60 per cent. Converting to pulp drying has another advantage if there is a market nearby, because dry pulp brings a greater return than wet pulp. The other process change, i.e., a shift from cell-type to continuous diffusers, is integrally related to recirculation of screen and press water. This change reduces the BOD generated by about 10 per cent.

Specific cases, as well as general industry examples, can be cited. For example, one chemical plant over a fifteen-year period doubled the dollar value of production and at the same time reduced the BOD and phenolic waste loads per day by over 90 per cent. The waste reductions were accomplished 'primarily by process modification, in many cases by simple changes such as increasing the temperature at which a reaction was taking place and changing a polymerization catalyst. In the production of potato flakes, the processing of raw potatoes by steam peeling results in a BOD waste with a population equivalent of about 200 per ton in contrast to about 420 per ton using caustic peeling. Where cooling is a major component of an industrial process, changes in cooling methods are changes in production processes. The adoption of air cooling, as in some chemical plants and petroleum refineries, will reduce the thermal waste load generated in the production process.

It is interesting to note that air cooling may be the rational economic choice even where water is relatively inexpensive in terms of both supply and waste disposal problems. The cost associated with circulating a large quantity of air must be compared with the costs of high head pumps, pumping energy, and pipelines required to circulate water, plus the costs associated with water intake and wastewater disposal. There is obviously no waste to be disposed of from air cooling units. In addition to the lower investment costs with air cooling, there are operating savings from less equipment maintenance and improved continuity of operation.

Table 4.3. Wastes Generated in Beet Sugar Plants, by Process 1 (pounds of BOD per ton of beets processed)

Operation	Potential BOD generation	After process changes	After recirculation associated with process change
Flume water	4.5	4.5	4.5
Screen water	2.52	2.53	0
Press water	2.6	2.63	0
Silo drainage	12.3	04	0
Lime cake slurry	6.5	6.5	6.5
Condenser water	0.7	0.7	0.7
Steffens waste	10.4	05	0
Total	39.5	16.8	11.7

In many cases it is difficult to separate changes in production processes from changes in products. Change in consumer demand over time is one stimulus for development of new production processes. Several examples illustrate the nature of product changes, with their implications for changes in wastes generation. In petroleum refining the product mix has shifted significantly over time toward the production of more gasoline per barrel of crude throughput. Continued changes in product mix are in prospect. For example, by 1995 the demand for jet-fuel will require about 16 per cent of all crude runs in the United States, in contrast to only about 3 per cent in 1965. This will require additional use of hydro cracking with consequent changes in waste generation per barrel.

In pulp and paper manufacture, there has been a steady shift toward a wider variety of products, such as colored paper and coated products of various types. Many of the coatings and sizings which have been developed to provide particular characteristics in the final products make reuse of the waste fiber more difficult or impossible, with a consequent increase in the waste load generated per ton of product. There also has been an increase in the proportion of pulp output which is bleached, from about 25 per cent in 1970 to almost

40 per cent in 1995. Bleaching increases the waste load generated in pounds of BOD per ton by about 25 per cent.

An example of a significant change in waste load through product change is the shift from hard detergents to soft detergents which are readily degraded by the biota in natural watercourses and in waste treatment plants.12 The change took place in Germany in October, 1994, and in the United States in the summer of 1995. Even so, soft detergents are still "under fire" because of their phosphate content, which contributes to algae growth in streams and lakes, and research is now oriented toward finding a substitute for phosphate.

Process and product changes have obviously had different effects on wastes generation. However, the dominant tendency has been toward a reduction in wastes generated per unit of raw or finished product. Two points are important: (1) most of the changes in production processes and product mixes have occurred independently of problems of water quality management (and of air and solid wastes management); (2) there are usually several ways at least of producing a given product mix, so that there are possibilities for modifications of production processes to reduce the wastes generated per unit.

One other type of in-plant change that can reduce waste generation is in-plant recirculation of water. As the extent of water recirculation in a plant increases, a decrease may occur in the waste load generated per unit, e.g., pounds of BOD and/or suspended solids per ton. This has occurred in the beet sugar industry and in the canning of some fruits. In the latter, where flume water is recirculated, less sugar is dissolved from the fruit in processing, thereby decreasing the BOD load per ton.

The separation of the effect of water recirculation *per se* on waste generation from the effect of changes in production processes and from materials recovery is not clearcut. In-plant recirculation of water may require modifications of production processes. Conversely, some production processes enable recirculation of water, as noted in the previous description of process changes in the manufacture of beet sugar. It is clear, however, that water recirculation in some cases has significantly reduced waste generation per unit.

Methods for Reducing Wastes After Generation

Materials recovery and by-product production are two methods for reducing wastes after generation. They form a logical pair, because the types of processes are often very similar. We differentiate between them on the basis of the final destination of their outputs. In materials recovery the output is reused within the same production unit as an input to the production process. By-product production yields consumption goods or intermediate goods used in other production processes.

Examples of materials recovery are to be found in almost all, if not all, of the heavy water-using or water-polluting industries. Only a few will be mentioned here for illustration. The recovery of chemicals in black liquor recovery systems in the production of sulfate pulp and the utilization of save-alls in paper production have been mentioned previously. In the manufacture of synthetic phenol by the sulfonation process, liquid wastes have been essentially eliminated by process engineering, and the value of recovered materials is reported to exceed recovery costs. In steel production, sedimentation to eliminate mill scale from waste discharges has resulted in the recovery of iron that can be used in the production process. Plating compounds, a potentially troublesome waste load, can be recovered from rinse water by evaporation. This procedure concentrates the rinse water sufficiently for reuse in the plating bath, which not only conserves valuable plating compounds but also significantly reduces waste and thereby the outlay for neutralizing chemicals. One plant anticipates an annual net saving of about $100,000 from use of the procedure. Waste liquor from citrus peel processing, normally a waste disposal headache, is concentrated to syrup and returned to the process. This procedure saves fuel as well as reducing the waste.

A number of industries have achieved substantial reductions in their final waste loads through by-product production. In beet sugar production, the recovery of monosodium glutamate and potash from the Steffens waste has significantly reduced the BOD generated. In canning apples, the utilization of the "waste" segments of the apple for vinegar production has reduced the waste load per ton. Cottage cheese whey has been converted into protein food supplements, thereby eliminating a difficult waste problem. Some of the by-products from waste materials in food processing are quite imaginative. For example, bran waste from milling has been used to make inexpensive sets of plates, casseroles, and trays, and coffee grounds have been used to produce discs and shelves.

As indicated, many alternatives are available for the reduction of wastes after generation by materials recovery and by-product production. In many cases these methods result in net profits to the industrial operation, even without counting the broader social costs avoided by the reduction or elimination of wastes discharged to watercourses.

After economic processes of materials recovery, by-product production, and effluent reuse have been exhausted, there are often residual waste loads which may have to be reduced or otherwise handled prior to final discharge. Some common methods for treating industrial wastes, also applicable to municipal wastes, are listed below:

Screening
Flocculation
Chemical coagulation
Flotation
Sedimentation
Centrifuging
Filtration
Stripping

Neutralization
Chemical oxidation
Chemical reduction
Wet oxidation
Fermentation
Emulsion breaking
Evaporation
Distillation

Incineration
Biological filtration
Activated sludge
Anaerobic digestion
Stabilization lagoons
Spray irrigation
Disinfection

In addition, the waste loads from some industrial operations may contain substances requiring special treatment, Such as colors, odors, special chemicals, and toxic materials.

The waste treatment methods listed above might be characterized broadly as physical, chemical, and biological processes. They have greater or lesser applicability to the two major types of wastes, nondegradable and degradable. With respect to the former, suspended and dissolved solids are the major wastes. Suspended solids can be removed by sedimentation, with or without the aid of flocculants such as polyelectrolytes, by filtration through various kinds of screens hind filters, and by centrifuging. Dissolved solids can be removed, to any desired degree, by one or more of the processes of distillation, ion exchange, electrodialysis, and reverse osmosis. It is possible to obtain completely pure water from waste discharges, but there will still be some sort of residual waste, either a concentrated brine or a semi-solid sludge. Ultimate disposal still remains a problem.

The degradable wastes of primary concern ire organic wastes, bacteria, and thermal discharges. Organic wastes are removed by physical processes, such as sedimentation, to some degree, and by chemical and biological processes, such as chemical oxidation and aerobic digestion, respectively. Bacteria are at least partially removed by biological processes and by disinfection. Thermal loads are removed by one or more types of air or water cooling systems.

Conventional treatment of domestic wastes consists basically of physical and biological processes, and is a comparatively standardized process. It can greatly reduce BOD, but will not completely, eliminate it, and is frequently designed to improve bacteriological quality and essentially eliminate suspended solids. Ordinarily the treatment of degradable organic wastes either domestic or industrial begins with the removal of the larger suspended solids by screening and grit chambers and the more finely divided suspended solids by sedimentation. New practices, such as the addition of polyelectrolytes have done much to improve the efficiency of the sedimentation process. This sedimentation stage, known as primary treatment, results in a wet, difficult-to-handle sludge, which is usually digested in heated anaerobic tanks before final disposal. In some cases the sludge is burned, usually after some degree of drying, or concentrated with centrifuges, prior to disposal. When waste treatment plants are operated adequately, primary treatment of an effluent containing an organic waste, with sludge disposal, can reduce the first stage BOD by 35 to 40 per cent.

What is generally termed a secondary stage of treatment is biological in character and essentially controls and accelerates the oxidation processes which occur in natural waters. Two major techniques are presently used, with various modifications. One is the trickling filter, in which a biological film is grown on rocks or some type of plastic medium, and the waste is applied intermittently, allowing air contact with at least the surface of the film. The other technique is the activated sludge process, in which air or oxygen is forced into a tank containing a mix of waste and actively feeding biota. Part of the settled sludge containing the biota is recirculated and mixed with the entering waste in the aeration chamber.

Each of these methods has certain advantages. Trickling filters require relatively little attention during operation, and the biological growths upon which the effectiveness of the filters depends are somewhat less susceptible to toxic substances than the activated sludge process. An activated sludge plant is somewhat more flexible in operation and design; the major variables are the proportion of sludge recirculated, the length of aeration time, and the amount of air or oxygen introduced. Rapid turnover of waste and sludge generally saves space and power, but requires well-trained personnel and continuous monitoring of plant performance. Treatment in all activated sludge plants can be made more effective by increasing the quantity of air passed through the sludge tank, although there is a definite limit to this procedure. Costs in activated sludge and other types of treatment plants increase rapidly as the removal of BOD is pushed above 95 per cent.

Small package plants, called extended aeration plants, have been used with considerable success for handling relatively small volumes of wastes, usually from strictly domestic sources. These plants are easy to operate, the sludge withdrawals are small enough in volume to be discharged with the final effluent, and much more of the solid matter is in mineralized form.

When secondary treatment is undertaken, the supernatant liquor from the sludge digestor, or other liquid resulting from the handling of sludge, is routed through the secondary treatment process. Secondary treatment in turn gives rise to some additional sludge which is routed into the digestors.

Primary and secondary treatment combined usually can reduce BOD by between 80 and 95 per cent. The procedures outlined above, plus disinfection of the final effluent in some instances, have been the essential elements of the primary-secondary waste treatment sequence for more than fifty years.

Although the conventional treatment processes are effective in reducing BOD, they have been much less effective in removing plant nutrients, nitrogen and phosphorus. During secondary treatment the nitrogen content of the organic waste is successively converted to nitrites and nitrates. Over-all, the volume of potential plant nutrients is reduced modestly during secondary treatment, but the content of nitrogen in nitrate form of course increases. Within limits, the extent to which nitrification occurs in the treatment process rather than in the receiving water can be controlled.

Other treatment methods that have become increasingly common and that achieve waste reduction comparable to secondary treatment are oxidation ponds (or stabilization lagoons) and spray irrigation systems.

Both methods are used in the food processing and pulp and paper industries. Ponds are used by the petroleum refining industry, by some large metropolitan areas (Melbourne, Australia, for example), and increasingly by smaller communities. Ponds, lagoons, and spray irrigation are used alone as complete treatment processes or as secondary or tertiary treatment in combination with other treatment processes. Ponds and spray irrigation systems may also be used in series, as in a number of installations in the food processing industry.

Stabilization lagoon or oxidation pond systems in industrial applications may range from a single large pond to a series of as many as eight or nine ponds. Pond capacity may be sufficient to store the entire waste effluent from a plant with a seasonal production schedule, as in some beet sugar and canning installations, or during the low streamflow period for plants with continuous production schedules. Alternatively, the pond system may be operated to produce a continuous outflow, as occurs in some installations in the canning, petroleum refining, and pulp and paper industries. A series of ponds may also have aeration devices of one kind or another between ponds and or floating or fixed aeration devices on or in one or more of the ponds themselves.

Algae play an important role in ponds and lagoons, as they furnish a substantial share of the oxygen requirements of aerobic bacteria, and they utilize the gases from the anaerobic BOD removal in the bottom sludge banks of such ponds and reduce or prevent their escape to the atmosphere. Such ponds and lagoons, when properly designed and loaded, are capable of stablizing oxygen-demanding wastes to a high degree.

In the Ruhr industrial area of West Germany, where comparatively large, shallow impoundments have been constructed in several rivers to serve essentially as large oxidation ponds in the streams themselves, the impoundments are sometimes specially designed to handle the specific types of wastes they receive. One, for example, neutralizes acids and precipitates organic materials simultaneously. One effect of the use of ponds, lagoons, and spray irrigation systems (and underground disposal) is the increase in the consumptive use of water per unit of raw product or final product, compared with conventional treatment. Data from the petroleum refining industry show that the mean consumptive use was nearly 50 per cent greater for refineries using ponds and underground disposal than for those using conventional treatment. Although the increase was small in absolute terms about 14 gallons per barrel of crude throughput it could be significant during periods of low streamflow, as in the summer months when the consumptive use in waste treatment by these methods and demands for water are both at a maximum.

It is important to note that both the oxidation pond or stabilization lagoon and the extended iteration treatment methods differ from standard waste treatment in that their effluents contain virtually all of the plant nutrients originally in the waste discharge. In the ponds, they are embodied in algae; in the latter, in mineral form. Thus the effluents from these types of treatment processes fertilize receiving waters more than effluents from conventional treatment plants do. Considerable attention is being devoted to developing methods for removal of plant nutrients from final effluents. Some recent results suggest that progress is being made on this problem. If fish or algae were harvested from oxidation ponds, the process of biological treatment would be accelerated and a major portion of the plant nutrients, as well as some other chemicals, would he extracted from the waste water.

Chemical treatment includes a variety of processes ranging from chemical oxidation of substances in waste discharges to neutralization of acid or alkaline wastes. Most types of chemical treatment involve low capital costs and high operating costs; this makes them particularly useful for specialized treatment problems, for final finishing of a sequence of waste treatment, and for infrequent application during critical periods of water quality. Some laboratory experiments with chemical treatment have shown a reduction of 60 to 85 per cent in BOD without biological treatment. Disinfection, by use of chlorine or ozone, can also be considered a type of chemical treatment.

One final point should be made concerning waste treatment. The increasingly stringent quality standards of final effluents make the economics of in-plant water recirculation more favorable. In some cases, water may be suitable for reuse within a plant even as process water-and yet not meet the quality standards for effluents.

The other major type of degradable wastes consists of thermal loads, which are of increasing importance. The most common method of reducing the heat load discharged is the installation of cooling towers. The magnitude of heat load reduction possible by this process is exemplified by the installation at the Philadelphia refinery of the Atlantic Refining Company, where the cooling tower program resulted in an 85 per cent reduction of heat load discharged to the river.

Whether or not a cooling tower is installed at a particular plant is a function of both the internal economics of over-all water utilization in the plant and of the offsite costs resulting from the discharge of thermal loads. With respect to cooling towers, the economics of recirculation are such that if recirculation is adopted, it will be done essentially completely. As a result, little or no heat will be discharged, but there will be an increase in the discharge of dissolved solids, stemming from chemical treatment of the recirculated water and concentration from evaporation.

Cooling towers may also aid in reducing other waste loads. For example, in the Toledo refinery of Sun Oil Company, a cooling water stream containing phenols passes through a cooling tower which functions essentially as a trickling filter. Biological oxidation reduces the phenol content in the effluent by over 99 per cent. Other applications of the same procedure have been reported.

A final method for handling residual wastes is the reuse of a municipal or industrial effluent by a subsequent user or users, usually after one or more types of man-made or natural treatment. Sewage effluent is used for various purposes in industrial operations, particularly in the southwestern United States. The practice is less common in the East, although effluent from the Baltimore municipal treatment plant is used in the Sparrows Point plant of Bethlehem Steel, primarily for cooling water in steel production. Effluents from municipal waste treatment plants have long been used for irrigation. Among the many cities that use all or part of their sewage effluents for irrigation are Grand Canyon and Tucson in Arizona; Bakersfield,

Fresno, San Bernardino, and San Francisco in California; and over 200 towns in Texas including Abilene, Kingsville, Lubbock, Midlands, San Angelo, and San Antonio.

Different reuse methods have varying effects on residual waste loads. Irrigation tends to remove virtually all of the residual wastes from the effluent, while industrial reuse may or may not reduce the residual waste load. Irrigation use, however, results in considerable net depletion of water, which is not necessarily unproductive.

Artificial recharge of groundwater aquifers represents another method of effluent reuse. In this case, however, generally a larger number of intermediate processes are involved than in the cases mentioned previously. Usually the effluent from a municipal waste treatment plant is inserted into a groundwater basin by means of injection wells or spreading grounds. In the latter case, the wastewater undergoes additional purification as it passes through the unsaturated zone. The purification process in the soil results from the action of soil bacteria, filtration, adsorption, and perhaps other processes. In other cases the waste effluent from the sewer system is conveyed directly to a wastewater reclamation plant, as at Whittier Narrows, California, where it is treated and then used for groundwater recharge. Another extensive system has been in operation for several years at Santee, California, where treated effluent is discharged into a series of ponds, the final ones having now been approved for water-contact sports. The final outflow is recharged to the groundwater basin.

Caution is necessary in using wastewater from domestic and/ or industrial operations for groundwater recharge. Even with treatment the minerals in the effluent before recharge remain, so that the total dissolved solids content is likely to increase (although there may be wide dispersion in the groundwater aquifer). Further, because there is likely to be some residual load even after treatment and the process of groundwater recharge, it is possible for groundwater contamination to occur. However, the well-publicized quality degradation of groundwater resources in certain areas, such as Long Island and Minneapolis-St. Paul, happened in situations where there was no centrally designed and operated wastewater treatment and recharge facility. Recent analysis suggests that groundwater recharge for water renovation may have much wider applicability and be more economical than has previously been appreciated.

An important point in connection with all measures of handling wastes is the existence of economies of scale. In materials recovery this is illustrated by the costs of the Blaw-Knox Ruthner sulfuric acid recovery process for handling waste pickle liquor. On the basis of 8.5 per cent free acid in the liquor, costs are about 3.5 cents per gallon when 30,000 gallons per day are processed and about 3 mills per gallon when 160,000 gallons per day are processed. Similarly, significant economies of scale exist in waste treatment. For example, in a study of fruit and vegetable canneries, the costs (annual charge on capital investment plus operation and maintenance costs) for lagoons and/or spray irrigation systems were found to vary from about 2 cents per pound of BOD removed for operations removing more than 500,000 pounds of BOD in a canning season to 30 cents per pound for operations removing less than 100,000 pounds. Extensive analyses by the U.S. Public Health Service show similar economies of scale for municipal waste treatment plants. In densely developed areas the costs of collective facilities for conveying and treating municipal and industrial wastes can often be much less than the sum of the costs of separate facilities for each municipality and industrial plant.

Reducing Waste Discharges

The foregoing discussion has focused on the wide variety of methods by which the quality of receiving waters can be improved by reducing waste loads discharged to them. The various methods can be combined in a myriad of ways to meet the needs of a single industrial operation or urban area or a complex consisting of many operations and or areas.

Consider just waste treatment to reduce waste loads. Even in a single industry the combination of treatment processes applied in different plants varies widely, primarily as it function of the raw materials utilized, the production processes involved, the final products, and environmental constraints, including effluent controls. Consequently, there is a wide range in waste treatment costs in any given industry.

What can be done to reduce waste loads discharged to receiving waters is clearly exemplified by the performance of the beet sugar industry. Table 4.4 shows the decrease in unit waste load (pounds of BOD per ton of beets processed) discharged daily during the production season by the industry between 1979 and 1992, and the measures adopted to bring about the decrease.

The types of measures to reduce waste loads used by the beet sugar industry are likely to be integral components of optimal water quality management systems. The planning of such systems requires the development of relationships between the incremental quantity of wastes removed and the incremental cost of removal. The relationship applies to a plant which has already installed pulp drying equipment and a continuous diffuser to which it recirculates pulp screen and press water, and which makes use of its Steffens residual. The costs shown relate to combinations of in-plant water recirculation and waste treatment.

Increasing the Assimilative Capacity of Receiving Waters

The inverse relationship between the concentration of most water quality variables and water quantity, providing water for dilution by fuller use of available streamflow through flow regulation suggests itself as a possibility. Low streamflows often coincide with heavy concentrations of waste loads, as well as with high temperatures. Where the waste loads consist of oxygen-demanding wastes, the dissolved oxygen level of receiving waters is thereby depressed. Periods of low flow and potentially large reductions in water quality are the conditions for which effluent treatment plants and effluent redistribution systems are generally planned and designed. The assimilative capacity of tidal estuaries may also be significantly affected by variations in inflow; but the effects are more dispersed and difficult to analyze than in the case of flowing streams.

Dilution of wastes beyond that normally provided by natural streamflow during low-flow periods can be provided by modifying the pattern of streamflow, by modifying the time pattern and 'or locations of waste discharge, or by some combination of the two. The most common practices are to increase flows during low-flow periods by controlled releases from reservoir storage, and to modify

Table 4.4. Estimated Average Reduction of BOD in Beet Sugar Processing
(pounds of BOD per ton of beets processed)

Type of waste	1979					1992				
		BOD removed					BOD removed			
	BOD generated	By process changes and recirculation	By waste treatment	Total	BOD discharged	BOD generated	By process changes and recirculation	By waste treatment	Total	BOD discharged
Flume and washer water	4.5	n.a.	n.a.	0.9	3.6	4.5	n.a.	n.a.	1.7	2.8
Cooling water and condensate	0.7	n.a.	n.a.	0.1	0.6	0.7	n.a.	n.a.	0.2	0.5
Pulp screen and press water	4.9	0.5	0.6	1.1	3.8	5.3	4.0	0.4	4.4	0.9
Silo drainage	12.3	5.8	1.2	7.1	5.2	12.3	12.1	0.1	12.2	0.1
Lime cake slurry	6.5	0.	3.1	3.1	3.4	6.5	0	6.1	6.1	0.4
Steffens filtrate	5.4	1.2	0.7	2.1	3.3	4.9	3.5	1.0	4.5	0.3
Total BOD	34.3	n.a.	n.a.	14.4	19.9	34.2	n.a.	n.a.	29.1	5.0

the pattern of waste discharge by temporary storage of wastes. Flow can also be increased by withdrawing water from groundwater sources and releasing it into surface watercourses. An interesting variation of low-flow augmentation is the recently inaugurated Buffalo River Improvement Project, in which water is pumped from Lake Erie, used as cooling water on a oncethrough basis in five industrial operations, then discharged into the Buffalo River to maintain a minimum flow.

The effectiveness of augmenting flow depends on both the type of waste and the type of receiving water involved. With respect to organic wastes, for example, increasing assimilative capacity by flow augmentation is effective for streams, but not for lakes, and only to it limited extent for estuaries. Increased flows could actually reduce water quality in lakes and estuaries by carrying greater quantities of partially assimilated wastes into them. With respect to nutrients, increased flows would likely result only in a more rapid discharge of these substances into lakes and estuaries, because less time would be available for adsorption and sedimentation in stream channels. In contrast, with respect to nondegradable wastes such as chlorides, increasing, the flow into estuaries can result in a significant improvement in water quality. It appears difficult, if not impossible, to increase the assimilative capacity of groundwater bodies by augmenting inflows.

Flow augmentation may play a large role in future water resources management in the United States. Large-scale planning efforts of the U.S. Corps of Engineers in several basins in the East will include flow augmentation for water quality improvement its an important feature. The final report of the Corps on the Potomac River Basin recommended a large amount of reservoir storage for increasing low flows to improve water quality.

The effects of reservoir storage itself on the quality of water later released may be favorable or unfavorable. On the positive side, bacteriological quality tends to be stabilized, and summertime releases from reservoirs tend to be cooler than normal streamflows. On the other hand, some early hopes for effective dilution through releases from reservoirs were disappointed because of the failure to consider the impact of impoundment on the dissolved oxygen content of releases. Water from deeper parts of reservoirs is often virtually devoid of oxygen, because of the combined effect of BOD demand

and reservoir stratification. When this condition is present as it is especially likely to be in areas experiencing hot summer seasons it takes larger releases from reservoirs to achieve it given effect on dissolved oxygen downstream, and the costs of achieving the desired water quality level are increased.

There are several ways of coping with the unfavorable effects of water storage in reservoirs. Multiple outlets can be installed as at the Oroville Dam storage unit on the Feather River in California so that water can be released from different levels in various combinations to achieve the desired water quality. If there is a power installation at a reservoir, air or oxygen can be introduced into water in the turbines. In many cases this can be done rather simply by using vacuum breakers already installed in the dam. Turbine aeration involves some power loss, and therefore is not costless.

Reservoirs can have significant effects on other elements of water quality such as suspended sediment, turbidity, and certain chemical constituents. An in-stream reservoir acts as a settling basin, and will eliminate or reduce problems of suspended sediment or turbidity. The clearer water may mean lower treatment costs for downstream water users, or it may promote increased algae growth with associated taste and odor difficulties.

Another method for modifying water quality in impoundments (and in natural lakes) is by mixing. The purpose is to destratify the impoundment or to prevent stratification from occurring. A number of installations have been made in the last ten years. Results show that it is possible to destratify an impoundment by mixing and to improve the quality of the entire water mass, but they also show that stratification can reoccur after mixing ceases. Hence, prevention of stratification appears to be the better approach. The desired circulation patterns can be achieved through reservoir design its shape and location of inlets and outlets as well as by mechanical mixing.

The degree to which the assimilative capacity downstream from an impoundment is improved is dependent not only on the reactions which take place in the Impoundment but also on how the reservoir is operated. Depending on the time pattern of releases, there can be favorable or unfavorable effects on the assimilative capacity. For example, when a power plant at a reservoir is used for "peaking," the release pattern will vary substantially. Flows will be high during the portion of each weekday when the peaking load is being met, and low during the night and on weekends. This pattern is in contrast to the more continuous pattern of waste discharges. Reregulation of the power releases can compensate for the varying flow, but of course at increased cost.

Another way of increasing the assimilative capacity of streams and lakes is by artificial reaeration with either air or oxygen. Mobile or fixed aerating devices can be installed in reaches of streams where needed to replenish the dissolved oxygen supply and prevent anaerobic conditions. Reaeration differs from methods for modifying wastes prior to discharge in that nothing is removed from the receiving water, and from methods for increasing the degree of dilution in that there is no increase in streamflow velocity to alter the length and shape of the oxygen sag. The effect is analogous to that of a lengthy oxidation pond artificially aerated.

In various coastal areas of the United States the heavy use of groundwater resources has led to saltwater intrusion. One solution to this problem is to reduce the pumpage from the affected groundwater aquifers. Another is to establish barriers to the intrusion. In Southern California, the Manhattan Beach Project for reducing seawater intrusion and preventing further deterioration of the available groundwater supply created a saltwater barrier by a series of injection wells. The well system provides a groundwater gradient toward the ocean which prevents further intrusion and enables the continued use of the groundwater aquifer. Impermeable physical barriers are another way of preventing saltwater intrusion.

Surface water is also subject to saltwater intrusion. Decreased outflow from upstream areas of rivers which empty into tidal estuaries has resulted in sporadic seawater intrusion, as in San Francisco Bay and the Delaware estuary. This type of intrusion can be reduced by increasing the flow into the estuary, or by constructing physical barriers to upstream saltwater movement. The first method has been utilized for many years in connection with San Francisco Bay, where the Central Valley Project has been operated to provide releases to reduce seawater intrusion. The second method has been proposed and studied for both San Francisco Bay and the Delaware estuary. The difficulty with physical barriers is that they impede navigation, affect aquatic life, and aggravate problems of managing wastes discharged upstream. Seawater encroachment in tidal estuaries will become more serious, as upstream development increases the consumptive use of water.

Effluent redistribution to make more efficient use of the available assimilative capacity can be achieved in a number of ways including regulated discharge, transfer of the effluent to a different location for discharge, and underground disposal. In general, these methods have little, if any, effect on waste loads as such, and they do not modify the receiving water.

Regulated discharge involves the temporary storage of waste effluents for discharge at times when more water is available for dilution. In an estuary, this may mean storage during the ebb tide and discharge on the flood tide. Often it involves storage of wastes during low-flow periods in the summer and discharge during winter and or spring when streamflow is higher. Some degradation of wastes takes place during seasonal storage, and this procedure is used by a number of foodprocessing, pulp and paper plants, and petroleum refineries. The efficient use of regulated discharge depends on sophisticated studies of the operation of such a system, and on a data collection system that provides continuous information on the quantity and quality of receiving water and the quantity and quality of wastes to be discharged.

Transferring of wastes from one location of potential discharge to one or more other areas where the assimilative capacity is larger is usually practiced in areas near tidal estuaries or oceans. This system is now used by the Toms River Chemical Corporation, which formerly handled the wastes generated in the production of synthetic and organic dyes by equalization, neutralization, clarification, chlorination, aeration, and discharge to the adjacent Toms River. When the residual waste load from growth in production approached the assimilative capacity of the river, ocean discharge was adopted. This made possible the elimination of both chlorination and aeration. The entire waste-handling system is continuously monitored.

In underground disposal, effluents are discharged into relatively deep formations that contain so little water or water of such poor quality that they are not expected to be used for water supply. This method of disposal is used largely by the chemical, petrochemical, and paper industries for wastes that are difficult or very expensive to treat. Underground disposal represents a consumptive use of water, since the effluent is not available for reuse.

Summary

The wide variety of measures for water quality management discussed above could be incorporated in plans for a water utilization system in a region (say a river basin). Planning, however, is only the first step in the management problem. The system must be implemented and once in existence must be operated to produce a specified time pattern of outputs of water and water-related goods and services, including water quality, at specified locations. This it must do in the face of uncertainties relating to hydrologic events and to demands on the water resource over time. Efficient and effective operation of a water utilization system involves not only day-to-day operation of individual units in the system, such as waste treatment plants and reiteration devices, but integrated operation of all units in the entire system. This is no mean task. Before concluding our comments on the technical and engineering aspects of water quality, we turn to a short discussion of system operation.

Operating a system means determining when to push buttons, turn valves, open and close gates, and so on for the various units involved in, and economic activities related to, it water utilization system. Such operation requires a communications-control network consisting of data collection, data communication, data analysis and interpretation, signaling, and response. The collection involves continuous recording of streamflow, precipitation, soil moisture, continuous monitoring of various water quality parameters in receiving waters and waste discharges, continuous recording of the volume and quality of water in storage reservoirs and in groundwater basins. In the last decade, considerable advances have been made in the development of instrumentation for continuous analysis of various water quality variables such his dissolved oxygen, temperature, turbidity, pH, total dissolved solids, conductance, and COD (chemical oxygen demand). From the data collection points the information must be transmitted to a control center. Again, recent developments improved this component of the management system through such techniques is automatic telemetering of data.

In the analysis and control center, the data are compiled and printed out or stored. Assuming basic relationships have already been developed, such as among streamflow, waste loads, and water quality, and streamflow and stream velocity, estimates can be made of water quality at various points in the system. This information is relevant to both shortrun and long-run situations. Examples of the former are accidental spills of toxic materials and intense storms producing high concentrations of suspended sediment which might clog water intakes or groundwater recharge facilities. Long-run situations involve protracted periods of low flow.

From the control center this information can be transmitted to various operators and water users in the area. The operators and water users at various locations can then respond to the "signals" received in terms of pushing buttons, turning valves, opening and closing gates. Artificial reaeration devices can be started; releases from reservoirs can be made; additional chemicals added in various treatment processes. Artificial groundwater recharge facilities may be by-passed. If necessary, generation of wastes can be reduced by curtailing production.

Despite all of the problems relating to system operation, the net gains which can he expected from the development of what we have called a communications-control network appear to be significant. Such networks were probably conceived first to preclude serious damage from accidental spills of toxic materials or substances that would disrupt waste treatment and 'or water treatment facilities (e.g., phenols, cyanide, arsenic) or to monitor the performance of waste treatment plants in relation to receiving waters. As demands on the water resource increase, with a concomitant increase iii interactions among various water users, the need for, and the potential savings from, sophisticated system operation will increase. Savings stem from the possibility for more nearly optimal system plans, i.e., reduction in size of some system components, and from the possibility of incremental operation of water quality management systems once they are in existence. The latter refers to the use at any given time of those measures which provide the desired water quality at least cost. The benefits from this type of system operation, including computer control, are well documented in the power industry.

It should be emphasized that the planning of any water management system involves assumptions about the way the system components are to be operated when they are finally In place. Accurate planning *requires* dctermination of the performance of proposed systems under various possible hydrologic events. Accordingly, to achieve a valid assessment of system performance requires structuring the simulation of the water utilization system in the planning phase in a manner as close as possible to the situation that will exist in the real world. Even so, the efficiency of operation of system components assumed in the planning phase can be approached, but never achieved, in actual system operation.

The degree of articulation and integration of water quality management system components implied in the above paragraphs has so far not been achieved even in system operation, much less in the planning of actual systems. The Ohio River Valley Water Sanitation Commission, however, has made substantial progress on the operation phase, especially with respect to a communications network. In most regions, water quality management is at a primitive level. At the same time, the technical and analytical means for implementing such systems are developing rapidly. Institutional means lag far behind.

Our discussion of the engineering and technological facets of water quality management has simplified many of the processes and relationships, but one central point should be clear. Technologically, there are many alternative measures available to achieve whatever levels of water quality ire desired. Those include various types and degrees of intake water treatment, measures for reducing the generation of wastes, measures for modifying and disposing of wastes after generation, and measures for increasing or making bet'er use of the assimilative capacity of receiving waters.

These measures substitute for one another economically at various rates depending upon the particular waste loads and receiving water bodies involved. Decisions concerning the proper balance among the various measures cannot be made rationally on technical grounds alone, because not all of the benefits from water quality improvement can be measured in dollar terms. Accordingly, values must be introduced into the decision-making process. Further complicating the decision process are both the stochastic nature (uncertainty of occurrence) of hydrologic phenomena and the inadequate understanding (as yet) of the impacts of waste discharges on water quality of various receiving waters and of the effect of various levels and time patterns of water quality on various water uses.

Identifying aternatives for a water quality management system and devising plans which combine them optimally is a demanding task. Moreover, planning and implementation are complicated by the fractionation of decision-making responsibility for the various possible components of such systems. Some measures are within the purview of the individual water users and waste dischargers, some are the responsibility of state, regional, or federal agencies, and some do not fall within the conventional purview of any agency whatsoever.

The problem is further compounded by the fact that water quality improvement is generally only one of the desired outputs from a water management system. Both system planning and operation are affected by the complementary and competitive relationships among the various outputs water quality improvement, energy generation, navigation, flood damage reduction, irrigation, water supply, and water-based recreation. We now turn to the economic aspects of regional water quality management.

Economic Concepts and Policies for Controlling Individual Waste Discharges

The limit on improving water quality in our regions is essentially an economic one. Although further technological advance may allow better quality with a smaller expenditure of resources, the controlling question will continue to be. "How much of society's resources shall we devote to maintaining and improving water quality?"

The primary aim is to develop certain basic ideas of economic theory in specific application to problems of water quality management. For the sake of simplicity it is assumed that no economies can be achieved by collective water supply and waste disposal facilities serving separate economic units such as industrial plants and municipalities.

The simplified situation presented in this part is one in which there are several water users along a stream. These water users are close enough together and the amount of dilution water is small enough for the costs of downstream users or the range of opportunities available to them to be affected significantly by upstream waste discharges. However, the water users are not close enough together to make treating their wastes in common treatment plants it feasible procedure. Also, technical or economic factors make it impractical to augment low streamflows or practice mechanical reaeration. The alternative ways of adjusting water quality, which *are* available in the region include all forms of treatment of water or wastewater it the individual water supply intake and wastewater outfall, temporary storage of wastes, alterations of manufacturing output, modification of industrial processes, temporary shut-down of industrial plants, and longer-run measures such as adjustments of industrial locations and changes in the location of industrial or municipal outfalls.

We made an effort to be reasonably comprehensive though not detailed in our earlier discussion of the various types of wastes and their effects, but we now simplify greatly and, in general, distinguish only between degradable and nondegradable wastes in developing principles and illustrating them. This distinction permits exploration of the relevant theory without undue complexity. Later when we examine measurement problems and introduce case material, we once more speak in terms of a fuller complement of individual wastes.

In general, this study is not concerned with the costs that must be incurred in moving water from a source to a final user or in draining a metropolis or industrial premises. Consideration of the problems of efficiency in planning and operating delivery and drainage systems is limited here to the suggestion that the relocation of points of intake and discharge may sometimes be an efficient alternative to other water quality control measures. The focus of this study is upon water quality in the streams or groundwater resources of a region and the optimum measures for its management. References to "watercourses" in the following chapters can usually be taken to mean either groundwater or surface waters. The same economic principles apply to groundwater and surface waters, but the type of analysis needed is affected by the fact that waste assimilation is the only "in-stream" use of groundwater, and the fact that quality deterioration is more difficult to reverse in groundwater.

It briefly explains why reliance on *private* incentives and competitive markets to allocate productive resources generally produce results that have *social* merit; it then points out why and how this allocative process may be distorted when watercourses are used for waste disposal. This form of market failure occurs because water quality deterioration produces "technological external diseconomies." These terms, traditionally used in economic theory. The definition of property rights in water and legal adversary proceedings are briefly evaluated as ways of handling external diseconomies. The possibility of enlarging firms so as to "internalize" the undesirable external effects and thus have them directly weighed in private decisions is also assessed.

The apparent deficiencies of these procedures lead to an examination of what might be called public administrative procedures-still focusing on individual points of waste generation and discharge. These procedures include effluent charges based on imposed damages and the more traditional procedure of effluent standards. Following an outline of the underlying analysis and knowledge of costs that are needed if effluent charges or standards are to achieve optimal control at individual points of waste discharge, attention is given to some of the more difficult problems encountered in making the requisite cost measurements.

❑❑❑

5

Waste Disposal Decisions and its Benefits

Introduction

The free market economy municipalities and private firms can escape certain costs associated with waste disposal by passing the problem along to other parties, and may find it to their economic advantage to do so. The ultimate costs throughout a whole watershed are not reduced in fact, they are increased if offsite costs exceed the costs of reducing waste discharges but they are borne by someone other than the dischargers. As we have already seen, the incentive to neglect offsite costs would disappear if a single competitive firm managed all phases of water supply and use throughout an entire basin, and if there were no public goods involved, because the external diseconomies would then become internal. The basin-wide firm idea, though useful for illuminating the character and role of external diseconomies, is not a satisfactory solution. Public intervention, however, can cause offsite costs to be reflected in the waste disposal decisions of individual firms and of local government units, and is needed for optimal resources allocation.

Some possible means of public intervention effluent charges, incentive payments, and enforcement of quality standards are considered in this chapter from the economic viewpoint. Discussion of political and administrative details of just what kind of organization might do the intervening.

In areas of considerable urban and industrial development the offsite costs associated with unregulated waste discharge are likely to be great enough to justify public regulation, despite the costs and distortions which may be entailed in the regulation itself. Here the focus is on the economic aspects of dealing with the external costs of waste discharge, and the assumption is made that regulatory public agencies exist and that they have sufficient geographical scope to internalize the major external costs associated with waste discharge in their areas. It is also assumed that the objective of these agencies, generally referred to as "river basin authorities," is to obtain the maximum net benefit from the water resources.

How can the basin authority induce the individual municipality or industry to consider offsite costs in its decisions involving waste generation, treatment, and discharge? This question is discussed, first, in terms of the method to be used. Emphasis is placed on a device whereby the basin authority would levy charges on effluents discharged. The alternative possibility of framing a system of payments for waste discharge reduction to achieve optimum water quality is also explored.

Attention is then directed to the variety of considerations involved in actually determining the costs associated with a given waste discharge. How can the effects of hydrologic variability be reflected in a system of effluent charges? Can the value of recreational, aesthetic, and public health uses be measured?

Economic Incentives for Reducing Waste Discharge

Economists have long held that technological spillovers can be counteracted by levying a tax on the unit responsible for the diseconomy and by paying a subsidy to the damaged party. Some have even demonstrated that under certain conditions the appropriate tax is just large enough to pay the appropriate subsidy. This idea makes a good bit of sense when one views the uneconomic effects of spillovers as resulting basically from a maldistribution of costs. However, in the case of waste loads, and from the point of view of resources allocation, it is not necessary both to levy a tax and pay a subsidy if the waste discharger and the damaged party do not themselves bargain about the externality. In principle, either a charge on effluents or a payment to reduce discharge will serve to induce the combination of measures that will minimize the costs associated with waste disposal in a region. But if bargaining takes place, a unilateral fee or subsidy will not produce an efficient result. To see this, assume that a public authority is created and directed to impose the cost of incremental damages upon the waste discharger. In other words, the authority endeavors to compute the relevant portion of a function like *YX′* in The horizontal axis measures units of waste withheld from discharge by the waste discharger and the *YX′* function indicates the incremental effluent charge which the waste disposer must pay for every unit of residual waste discharged. If damages are properly determined and assessed, does optimization require that the loss to the "damaged" party be compensated? If the authority were in fact to levy a charge equal to the damages associated with each increment of waste discharge, it would collect just enough in tolls to cover the residual damages (area *XCX′*).

From the viewpoint of efficiency (as contrasted with equity) compensation must be paid if the parties can and do negotiate. Otherwise, optimal resource allocation will be prevented both in the short run and the long run, because the level of waste reduction that minimizes the sum of party 1's cost for waste reduction plus party 2's residual damages does not necessarily minimize party 1's cost plus party 1's charge plus party 2's damage. Consequently, a schedule of charges equal to marginal damages imposed and not used to pay compensation will not eliminate what have been termed "Pareto-relevant" externalities at the optimum level of waste reduction, i.e., from the viewpoint of the parties further gains from trade are possible.

Function *AC* indicates the incremental damage costs imposed and *YX′* indicates the marginal gain to the waste discharger due to not having to reduce his waste discharge (say by treatment). If the marginal cost of offsite damages is levied upon the waste discharger, his net gain becomes function *YX*. At the optimum point *X*, his marginal gains are cancelled by the marginal damages imposed so that his net gain is zero. This is another way of saying that *X* is the optimum point. Acting by himself, if marginal offsite damage costs are imposed upon him, the waste discharger would tend to move to this point. An external effect continues to exist at the optimum point. This simply indicates that the benefit to the discharger from not reducing the waste load further is greater than the damage imposed on other parties.

Unless, however, the damaged party is compensated so that he experiences no damage costs at the rate of discharge indicated by point *X*, he will be willing to pay the waste discharger to reduce his waste discharge further to the point where the discharger's marginal net gain equals the marginal damage imposed.

If the parties can not or do not negotiate, the social optimum can be attained by taxing the waste discharger and not compensating the damaged party. We therefore conclude that achievement of an optimum by means of effluent charges would not in general involve compensation of parties adversely affected by waste discharges. In the remainder of the book we assume that the parties involved cannot negotiate with each other and that no compensation is paid. The discussion up to this point has been conducted in terms of effluent charges representing incremental offsite or external costs imposed. It is not difficult to demonstrate that, *in principle*, the same result could be achieved by making payments for the reduction of waste discharge-even under conditions where longer-run adjustments such as plant location and process change are involved. The distribution of income would of course be different under the two arrangements.

Payments

We believe that a system of payments, or "bribes" as they have recently been termed in the literature, could in principle achieve the same result as an optimal charges scheme, despite some recent statements to the contrary.

Assume that a profit-maximizing firm has an incremental production cost curve as indicated by *MC*, that the price at which the firm can sell the commodity it produces is given, as indicated by the curve *D*, that the *only* way the firm can diminish the amount of a residual substance which it discharges into a stream is to reduce production, and that residual waste per unit of output is a constant. If a regulatory authority imposes a unit charge on the effluent of the firm, the incremental production cost function will shift upward by the amount of the charge per unit of output (i.e., charge per unit of effluent times effluent per unit of output) from *MC to MC′*. On the other hand, if the regulatory authority offers to pay the same amount per unit for reducing waste discharge, the incremental cost function will still be *MC′*. A firm rationally trying to maximize its profits will view the payment as an opportunity cost of production because waste discharge is, by assumption, a straightforward function of production. This proposition can perhaps be clarified by the numerical example below:

Output of product	Total cost of production	Incremental cost of production	Fffluent charge on waste discharge per unit of production	Payment for reduction of waste discharge per unit of production	Minimum product price necessary to induce indicated no. of units of production with charge or payment
0	10	...	...	...	$...
1	11	1	1	1	2
2	13	2	1	1	3
3	16	3	1	1	4
4	20	4	1	1	5
5	25	5	1	1	6

If, for example, the incremental cost of producing the fifth unit (neglecting the cost of waste discharge) is $5.00, and a public authority stands ready to pay $1.00 if the manufacturer does not discharge the waste associated with that unit, he will produce that unit only if the price is at least $6.00. If the price were less than $6.00, his net income would be higher if he simply accepted the $1.00 payment. The $1.00 payment thus becomes an *opportunity cost* even though it is not a direct outlay. The point is that in the sense of *opportunity cost*, which is the relevant concept of cost for decision purposes, the two procedures have entirely the same effect on incremental costs.

The question may well be raised whether in terms of longer-term adjustments, i.e., a firm's decision to enter or leave an industry or to expand or contract production capacity, the effect may not be different. The answer is in principle "no," but great informational and administrative difficulties emerge if the payments route is adopted. These are discussed later.

Assume that all plants in the industry are initially earning no "excess" return (i.e., they are earning only the minimum necessary to keep their resources employed in the industry). The new long-term equilibrium adjustment would then find the industry price higher by the amount of the subsidy or the payment, fewer firms in the industry, and fewer resources devoted to producing the output of the industry. All this would reflect the fact that costs imposed elsewhere in the system, and previously neglected by management, have been internalized to individual firms.

A system of charges or payments reflecting downstream costs would, however, probably not raise the opportunity costs of all firms by the same amount. Under these circumstances and assuming costs were otherwise the same the plants inducing the highest offsite costs would be selectively shut down. Plants making an "excess" return, would not be shut down, but their output would be reduced, and prices would tend to rise to reflect the opportunity cost of production.

The effects of waste disposal costs on prices, location, and decisions to enter or leave an industry will usually be small, because these costs are generally quite low relative to other costs of production and because the firm has the option of treating or otherwise modifying final waste output.

Decisions on location and/or industry entry or exit, however, have a major effect on the administration of a system of payments. Payment must be continued after shutdown of a plant, if the procedure is to have the desired results. While this might be manageable, serious problems would occur if a shift in demand for the product should increase the potential profitability of the plant, or if other dynamic adjustments should take place. Moreover, plants might introduce processes that produce a greater quantity of waste in order to obtain payment for reducing waste.

Even more perplexing would be the handling of proposals for new industrial locations. The administrative authority would have to stand ready to make payments to industrial plants which never do locate in the area but which would do so if a payment reflecting the costs their effluents would impose were not made. Payments on this basis would of course be an open invitation to extortion. If charges were levied, however, the authority would only have to provide the prospective firm with an estimate of the unit charge to be placed on its effluent.

To recapitulate, the basic point about the payments-charges controversy is that only the payments scheme requires information about a "status quo" point or rate of discharge. Obtaining such information would be difficult; if industries enter or leave, it might be impossible. Moreover, the payments technique would make it easy for the unscrupulous operator to benefit by exaggerating his potential waste load.

Thus far, it has been assumed, for the sake of simplicity, that the only way in which waste loads can be reduced is by decreasing production. Waste loads can, of course, be reduced by other means including industrial process changes, materials recovery, by-product production, and treatment. That a payment per unit reduction of waste load win produce the same balance between production cutback and treatment (or other methods) as a charge of equivalent amount can be conveniently shown by means of an isoquant diagram. The isoquants *CC′* and *KK′* indicate along their lengths a constant amount of waste reduction (say, pounds of BOD) which can be achieved by alternative combinations of treatment or production decrease.

The lines *AA′* and *BB′* indicate the number of units of treatment capacity and reduction in output that can be purchased at equal cost. In principal they could relate to any combinations of alternatives. In many industries, internal process adjustments will be economical substitutes for external treatment. If more than two alternatives are involved, diagrammatic presentation becomes cumbersome if not impossible, and mathematics must be used.

Point *A* shows the number of units of production which can be foregone at the expense of an amount of net revenue equivalent to the outlay necessary to reach *A′* (disregarding waste disposal costs). The line *AA′* represents all the alternative combinations of production foregone and outlay for increased treatment which can be obtained at the same cost. The line *BB′* represents a similar set of alternatives for a higher cost level. For simplicity these functions are pictured as straight lines although this is not essential to derive the optimization criteria outlined below. An infinite number of curves like *AA′* and *BB′* corresponding to all possible levels of cost could be drawn.

Expenditure on treatment and loss of revenue on foregone production of incremental units may be in the interest of the firm either to avoid a charge on effluent or to obtain a payment for reducing discharge. A central point is that for any level of cost (outlay or foregone revenue) the firm will seek that combination of measures which will achieve the greatest reduction in waste discharge. That combination is attained for the cost level *BB'* at point *X'* where the constant cost curve is tangent to the constant product curve *KK'*.

An infinite number of curves like *KK'* and *CC'* could also be drawn corresponding to each level of waste reduction. The convex-to-theorigin slope of these curves indicates a diminishing marginal rate of substitution between the two alternatives. As treatment capacity is expanded, it substitutes less well for output reduction. Another way of saying this is that there are diminishing physical returns to increased retention time in the treatment plant.

Points like *X* and *X'* indicate greatest effect at a given cost as no higher isoquant or constant product curve (in this case product refers to waste discharge reduction) is attainable given the combinations of input that can be purchased. At such points the combination of the two alternatives pictured is economically optimum the rate at which one can be "traded off" for the other at the margin in order to produce a given output is equal to the ratio of their respective prices, and we can conveniently refer to the cost of output reduction and of increased treatment input as their respective "prices." The rate of tradeoff or marginal technical rate of substitution is equal to the slope of the isoquant. It is also readily demonstrable that the slope of this curve equals the marginal physical product of treatment (i.e., the physical output result produced by a small increment of treatment) over the marginal physical output reduction. Since the slope of the price line (like *AA'*) equals the ratio of the prices of the two alternatives. A marginal cost function analogous to the one, but relating the marginal cost of an combination of waste reduction measures to the degree of waste reduction achieved can readily be constructed from the information contained in an isoquant diagram by establishing numerous points such as X and reading off corresponding costs and outputs. The optimal waste reduction level can then be established by equating the marginal cost of the optimum combination of output reduction and treatment (or other methods) with the charge or payment representing the downstream incremental cost. When this condition exists the marginal costs of all relevant alternatives (including treatment, process change, output reduction, and pollution damages) are equated and the level of waste production and treatment is optimized. The major point here is that since the costs of alternatives are evaluated equivalently under the charges and payments procedures, both the combination and level of use of alternatives will be the same.

In actuality, lines like *AA'* and *BB'* would frequently be curved. They would probably be convex to the origin, since with rising marginal costs the cost per unit of reducing output would tend to rise. (There is an important distinction here between a rising price and rising incremental costs.) Straight lines (given *P*'s) were assumed in order to simplify the derivation of the optimizing rules, and the demonstration that effluent charges and payments for waste reduction lead to the same optimizing rules. Conclusions with respect to charges and payments remain the same even if equal cost lines like *AA'* curve, provided certain constraints on their curvature are met.

The basic and quite reasonable rule remains that in order to minimize costs the marginal costs of all relevant ways of achieving a given result must be equalized. Moreover, the result will be the same whether the cost is an actual outlay or the foregone opportunity to receive a payment. This section has illustrated that, under idealized conditions, payments and charges serve equally well to achieve optimal amounts of waste discharge reduction for both the short-run or the long-run situation and for multiple alternatives to reduce waste as well as for output reduction.

But it has also illustrated the extensive informational and administrative requirements of the payments scheme. Accordingly, the discussion in the remainder of the book is almost entirely in terms of "putting the costs on the waste discharger." If costs are properly defined, this procedure tends to produce an optimal allocation of resources and optimal residual or final waste loads. However, water quality management is such a complex matter that any procedure is bound to be accompanied by problems of adjustment, equity, or inefficiency, or perhaps all three. The effluent charge procedure is no exception. It presents problems of adjustment and equity especially when locations have been made on the assumption that a certain amount of waste can be discharged freely.

Problems in Determining Charges

Basic to an optimum system of effluent charges, or for that matter any procedure aimed at achieving an economically optimal level of

Table 5.1. Simple Illustration of Damage Distribution

Plant no. (*serially located along stream*)	Chloride load dis-charged (*1,000 lb. per day*)	Chloride load at plant intake (*1,000 lb. per day*)	Flow Condition I: Streamflow (*mill. gpd*)	Flow Condition I: Chloride concen-tration (*1,000 lb. per mill. gpd*)	Flow Condition I: Damage per day (*$1,000*)	Flow Condition I: Total damage per day (*$1,000*)	Flow Condition II: Streamflow (*mill. gpd*)	Flow Condition II: Chloride concen-tration (*1,000 lb. per mill. gpd*)	Flow Condition II: Damage per day (*$1,000*)	Flow Condition II: Total damage per day (*$1,000*)
1	1.0									
2	0.5	1.0	1.0	1.0	1.00	1.00	0.5	2.0	2.00	2.00
3	1.5	1.5	1.0	1.5	3.00	4.00	0.5	3.0	6.00	8.00
4	1.0	3.0	2.0	1.5	3.00	7.00	1.0	3.0	6.00	14.00
5	0.5	4.0	2.0	2.0	1.00	8.00	1.0	4.0	2.00	16.00

Damages caused at	Damaged caused by:		Plant 1	Plant 2	Plant 3		Plant 4	Sum of damages caused to	Damage caused by:	Plant 1	Plant 2	Plant 3	Plant 4	Sum of damages caused to
Plant 1	0.00	0.00	0.00	0.00	0.00	0.00	0.00	0.00	0.00	0.00				
Plant 2	1.00	0.00	0.00	0.00	1.00	2.00	0.00	0.00	0.00	2.00				
Plant 3	2.00	1.00	0.00	0.00	3.00	4.00	2.00	0.00	0.00	6.00				
Plant 4	1.00	0.50	1.50	0.00	3.00	2.00	1.00	3.00	0.00	6.00				
Plant 5	0.25	0.125	0.375	0.25	1.00	0.50	0.25	0.75	0.50	2.00				
Sum of damages caused by:	4.25	1.625	1.875	0.25	8.00	8.50	3.25	3.75	0.50	16.00				

water quality, is delineation of the "damage cost function," which is the functional relationship between the amount of a waste discharged and damages.

In this early discussion we do not distinguish between short-run and long-run damage functions. This important distinction is introduced later. Also we initially ignore the stochastic character of the function resulting from environmental variability primarily temperature and streamflow. This permits us to first discuss basic principles without introducing duration and frequency as affecting the magnitude of damages.

The simplest situation occurs when this function is linear, i.e., each additional unit of waste discharge results in an equal increment of damage. In general, the present exposition will assume linear damage functions. Such a situation is illustrated in Table 5 where it is assumed that five plants are arrayed along a stream, that streamflow increases along the course of the stream (say, because a tributary enters), and that the waste it contains is nondegradable (say, chloride). The nondegradability assumption does not change the analysis in any way, but it simplifies the example.

Since damage per day is assumed to be in direct proportion to concentration, a level of charges equal to incremental damage costs can be worked out for each level of flow and for each plant. For example, at flow level one, the charge for plant 1 is $4.25 per pound of waste discharged which is the sum of the damages caused by plant 1 to plants 2, 3, 4, and 5. (At flow level two, the charge for plant 1 is $8.50 per pound, because the same waste discharge results in a doubling of the concentration.) The charge for plant I is the same regardless of the level of discharge of the other plants. If plant I reduced its discharge by half in response to the charge levied on it, its assessment would drop by half to $2.125 the amount by which downstream damages are reduced. This "separability" characteristic of linear damage functions is very important since it greatly reduces the amount of information which a regulatory authority must have to implement an efficient effluent charge system.

The manner in which a plant maximizing its profits (or minimizing its losses) will respond to an effluent charge levied on it at a given level of streamflow and the effect of this response on costs associated with its waste disposal. Under the circumstances pictured, the plant will reduce its waste discharge from point *D* to point *E*, thus minimizing the costs associated with its waste disposal. The marginal cost of withholding wastes in an optimal manner includes, in the case of chlorides, such alternatives as temporary storage, process adjustments, and reducing production. The marginal cost function rises after some point because it becomes progressively more expensive to withhold production (say, as inventory is run down), or to adjust production processes.

- *OD* = Units of waste discharge if no charge levied on effluent.
- *OA* = Damages per unit of waste discharge and effluent charge per unit.
- *OACD* = Total damages associated with unrestricted waste discharge, i.e., no effluent charge levied.
- *OE* = Reduction of waste discharge with effluent charge *OA*.
- *OFBE* = Total cost of reducing effluent discharge to *ED*.
- *OFBDC* = Total cost associated with waste disposal with *ED* waste discharge, i.e., residual damage costs plus cost of reducing discharge.
- *OABE* = Total damages avoided.
- *ABF* = Net reduction in waste disposal associated costs by reducing waste discharge by *OE*, i.e., *OABE* minus *OFBE*. This also equals the total cost of not reducing the effluent discharge.

The costs of increments of effluent storage capacity may rise if it is necessary to purchase higher-priced land for storage or if the land providing gravity flow to the stream is exhausted and some pumping costs are incurred. The firm acting to minimize its costs (charges plus costs of avoiding them) would continue to use each alternative until $MC_A = MC_B = MC_C = \ldots = MC_N =$ unit charge; in other words, until no further marginal "tradeoffs" are possible which will reduce costs. (In Part III problems of long-run planning are contrasted with short-run problems of adaption with existing capacity.)

The relative simplicity of a system of effluent charges based upon linear damage functions arises from the fact that each waste discharger's damage costs can be determined separately. In contrast, non-linearities are associated with interaction between waste discharges in the sense that the incremental external cost associated with one cannot be determined unless the level of discharge of the other is known. To put this in another way, there are cross product terms in the damage function. We refer back to the damage functions indicated on pages 94-96. where the "separability" case yielded a damage function of the form $C = K(K_1L_1 + K_2L_2 + K_3D_a)$ whereas the nonseparable function was $C = K(K_1L_1^2 + K_2L_2^2 + K_3L_1L_2 + K_4L_1D_a + K_5L_2D_a + K_6D_a^2)$. In this equation, the damages from the first discharge are made up of the first, third, and fourth terms. The third term depends upon the level of the second discharge and, therefore, the damages attributable to the first discharge cannot be determined without knowing the second discharge. The problem this presents is that instead of merely needing to know damage functions and being able to achieve an optimal solution by imposing damage costs on the waste dischargers and letting them respond, the agency must now know the cost functions for waste reduction at each interdependent point of waste discharge. It must determine the optimum level of waste reduction for each point *before* charges are assessed on effluents in order to obtain an optimal solution. This means that the effluent charges system loses perhaps its major advantage over other systems of control such as payments or effluent standards. This point is explored further below.

In some cases it might be preferable to assume linearity even when it is known that the discharge-damage relation is actually somewhat more complex because taking account of greater complexity may rapidly increase costs and yield strongly diminishing returns. In other words, among the incremental costs that should not be neglected are the costs of refining the system of charges.

Some Additional Theoretical Problems in Determining Optimal Charges

Even when damage functions are linear, "separability" problems may arise if waste disposers impose damages on a series of downstream units and the waste discharge of one firm affects the marginal production cost of another. A separable externality is defined as one that occurs when the total cost of a productive process is affected by the level of output in another process (and hence the level of upstream waste discharge), but marginal cost is not. The separability case requires that the marginal production cost of each firm linked to another by a technological externality be given entirely in terms of its own output. This would follow, for example, if factory A discharged a waste which caused factory B to put in a water supply treatment plant which thereafter would be operated in an invariant fashion. The new total cost function becomes a vertical displacement of the old and marginal costs are not changed, as is shown in the following example.

Plant B output	Total cost before factory A	Marginal cost	Total cost after factory A	Marginal Cost
1	1.0		2.0	
		1.5		1.5
2	2.5		3.5	
		2.0		2.0
3	4.5		5.5	

Since marginal costs of one firm are not changed by the output decisions of the other firm, the externality does not cause output to be different from what it would have been in the absence of the externality presuming it continues to be in the affected firm's interest to produce at all. The effect of the externality is intra-marginal and thus merely affects the firm's profit position and the decision whether to continue production or to stop. In this instance, the appropriate effluent fee is simply a lump sum equal to the increase in total cost. This is a fee levied for the privilege of discharging at all. The only question involved is whether to discharge or not.

A non-separable externality is defined to occur when the marginal cost in a productive process is affected by the level of output in another process. For example, assume that a petroleum refinery A, which expels a hot effluent, locates upstream from a petroleum refinery B, which is operated by a different firm and which uses the stream for cooling water. Cooling efficiency for plant B will drop, and for the purposes of our example we assume it will have to pump more water for cooling per unit of output. The marginal costs of plant B *for a given level of output* will be affected by the externality, shown on next page.

A particular marginal cost function for plant B corresponds to each level of discharge of plant A. Plant A can determine its own profit-maximizing level of output, and at this rate of production a specific level of output (say, x_1 corresponding to MC_1) will also maximize profits for plant B (although this will in general not be the level of output which will maximize joint profits for both plants taken together). However, it is possible to compute the amount of a levy on the output of plant A which will maximize joint profits. The appropriate level of charge will be the one which causes the marginal net loss imposed on firm B to be considered an opportunity cost in firm A. The net damage or net loss function may also turn out to be linear as indeed. But even so we are not out of the woods if different waste discharges and water users impose external costs on multiple production units.

Plant A discharge	Total pumping costs before plant A	Marginal costs	Total pumping cost after plant A	Marginal cost	Total external cost of plant A discharge	Marginal external cost
1	1.0		2.0		1	
		1.5		2.5		1
2	2.5		4.5		2	
		2.0		3.0		1
3	4.5		7.5		3	

Assume an initial equilibrium along a stream where a number of water users and waste dischargers impose external diseconomies on one another seriatim. Assume the regulatory authority starts by imposing the incremental costs of the last one of these (*Z*) upon the previous one (*Y*) and so on back up the line. When the charge is levied upon Y the outputs of both *Y* and *Z* change. As the cost imposed by *X* on *Y* is levied on X the output of *Y* changes again; accordingly so does that of *Z*. We have an interdependent system, and a simultaneous solution is necessary for the entire system. Not only is the problem analytically complex, but it would require detailed information on the costs of reducing waste discharges as well as damages imposed, and once again we lose the two major advantages of the effluent charges system as a means of water quality management. Decentralized decision making is no longer possible, and the information needed by the basin agency is no longer limited to damage functions.

Things may not be so dark, however. For one thing, municipal waste loads are not sensitive to damage costs imposed upon the municipality, and thus do not give rise to serial effects. Recreation, one of the major uses affected by quality deterioration, ordinarily does not itself produce a significant waste load. Moreover, some industrial damages such as corrosion appear to be separable. In other instances, effects on marginal costs may be too small to affect output appreciably. The basin agency may well find it a useful rule-of-thumb to assume linear damage functions and neglect serial effects. However, it must be aware that these are simplifications, the benefits and costs of which must be weighed against reality.

Probabilistic Character of Damages

The Expected Value of Damages. One aspect of damage measurement importantly related to the problem of levying appropriate effluent charges (as well as establishing standards) is the probabilistic character of damages. As already explained, the damaging effect of a given waste discharge is heavily dependent upon the natural environmental conditions which prevail when the discharge is made. Most important are conditions of water availability and temperature. Although records are often short, methods have been devised which permit analysis of the characteristics of frequency distributions of streamflow and temperature. In general, they rely on chain-like statistical sequences known as Markov processes. We take for granted that the statistical problems can be handled and proceed to discuss the effects of probability or risk elements on damage functions.

Let us direct our attention in the first instance to an individual outfall, at which a nondegradable waste is discharged. Assume that there are two water users downstream from the given outfall and that the damages they incur are directly proportional to the concentration of the waste in the raw water. For example, these users might have damage functions, D_1 and D_2, each of which represents the costs of the optimum combination of water supply treatment measures and residual physical damages, for each level of concentration. For each quality of raw water, the downstream users would, as we have seen previously, have an incentive to combine the alternatives open to them in this way. Because the functions are linear and separable, the total damage corresponding to any level of concentration is the sum of the damages to each user, i.e., $D_1 + D_2$.

Since, for a nondegradable waste, dilution is the only factor affecting concentration, at any given point of use the damages will vary with the streamflow. Since streamflow is a stochastic phenomenon, downstream damages are necessarily probabilistic variables.

Damages avoided by reducing waste discharge can be estimated by constructing a new cost-frequency curve which shows the probability of damages corresponding to various lower levels of waste discharge. For example, a flow of 300 mgd is expected to be equaled or exceeded 80 per cent of the time. Corresponding to that flow are damages of *X* dollars, resulting from a given quantity of waste discharge. If the quantity of waste discharge is reduced, but the flow regime remains unchanged, the damages associated with the *same* level of discharge are decreased, i.e., to *Y* dollars. To put it another way, the same level of damages, i.e., *X*, now corresponds to a flow which is equaled or exceeded only about 60 per cent of the time. Thus, the probability distribution of damages has been changed.

Taking the integrals of these functions (roughly, multiplying probabilities times costs) gives the expected value of damages. Then, taking successive differences between the integrals corresponding to different levels of waste discharge yields a series of expected values which can be used to construct a curve showing the relationships between diminished residual waste discharges and the reduction in the expected value of damages.

An effluent charge that does not vary over time should ideally reflect the *expected* value of damages avoided when waste discharge is reduced incrementally. The waste discharger would then have an economic incentive to plan and design his waste reduction facilities in a way that should minimize the over-all costs associated with his waste discharge, given that his waste reduction facilities are to operate at a constant level regardless of the variation in flow and changes in the concentration of waste at the points of downstream use.

However, neglecting for the moment the added costs of administering a more complex system of charges, this is not the minimum cost which could be obtained if the effluent charge reflected changing external (downstream) costs over time. If the charge did do this, the waste discharger would be induced to consider the expected value of the charge (damages) as he or the management agency calculate it in determining the capacity of his facility (a treatment plant, for example), and would *operate* the plant to reflect short-run changes in external costs. He would also consider the percentage of the time the plant would operate at various levels of waste reduction in designing it (i.e., in trading off capital costs against operation costs). Reason suggests and empirical research demonstrates that this approach can reduce over-all social costs below the level entailed in constant rate operation. This is the rationale for searching for ways of devising an effluent charge system which will reflect shorter-term variations in external cost.

Treatment Plant Design. To be a bit more concrete about the last point, we focus briefly on the matter of designing and operating a waste treatment plant in such a manner as to minimize cost. For simplicity we contemplate a situation where the objective is to maintain a given standard in the stream. This yields results that are somewhat different from those produced by a true equating of internal and external marginal costs. However, the simpler situation can serve to illustrate the point.

Plants are usually designed to achieve lowest cost at a particular level of waste removal predicated on a particular low streamflow of relatively infrequent occurrence, say 10 per cent of the time. More often than not, the plant is then operated at about that level regardless of changes in the stream's assimilative capacity. Long-term costs would tend to be lower if the plant were designed to operate most efficiently at a higher, more frequent flow. The added treatment required at low flows might be provided by adding chemicals, by increasing the degree of aeration or the amount of recycling, or by other means that entail high operating costs. Operating costs can be quite high during the rare periods of extremely low flow and still yield a comparative advantage over additional capital investment.

For example, let us assume that a plant is designed to deal with a lowflow condition that is likely to occur once in ten years. Let us further suppose that the probability of this event occurring in any given year is equal to that for any other year. The expected value of the extra cost of dealing with such an extreme low flow can then be treated as an annual average. It is possible to establish how high this annual average can be without its present worth exceeding a dollar of current investment by determining the present worth of a constant expenditure stream of $1.00 per year. The present value of such a perpetuity discounted at 5 per cent annual interest is $20.00. This means that 5 cents could be spent annually, or 50 cents during the once-in-ten-year-probability low flow, and the expected present value of such an expenditure stream would not exceed $1.00. In other words, 50 cents could be spent every 10 years in perpetuity, and the cost would probably be no greater than spending $1.00 for capital equipment now. This means, for example, that if $2 million of the capital value of a large plant is for the purpose of dealing with the once-in-ten-year condition, $1 million could instead be spent on operating costs every ten years in perpetuity at no greater expected cost.

Some empirical information is now available concerning the possible cost savings for variable rate operation. One study by Frankel explored the hypothesis that significant cost savings might be achieved in connection with a waste management system based upon streamflow conditions in the Eel River in California. Frankel first estimated the costs of achieving stream standards based upon continuous rate operation of conventional waste treatment plants. Next, by varying the degree of treatment to just maintain the standard, he found that operating costs could be reduced by about 15 per cent per year. He also explored the possibility of storing wastes during low flows and discharging at higher flows, and found that this produced a saving in costs of about 30 per cent over continuous rate operation of treatment plants to maintain the minimum standard.

The importance of tailoring waste removal operations to external costs, or to the requirements of a stream standard, will become greater when a high degree of treatment is required. In a study of high-level treatment alternatives to meet projected waste levels and desired water quality standards in the Potomac estuary, Robert K. Davis found that varying the level of operation of advanced treatment facilities could not only reduce cost greatly but might alter the cost relationships between methods. This result is illustrated for two types of advanced treatment techniques. Both techniques are able to meet the specified standard either by continuous operation or by an average of 3.5 months of intermittent operation per year.

A system in which effluent charges would be varied to reflect changing environmental conditions, i.e., taking into account that given waste loads impose different external costs over time, could be a powerful stimulant for the design of processes that can respond flexibly to changed conditions and for the optimal operation of such facilities.

Table 5.2. Present Value of Treatment Costs

Process	Continuous operation	3.5 months per year operation
Step aeration plus microstraining	$ 49,000,000	$39,000,000
Chemical precipitation	110,000,000	32,000,000

Note: Adjustments made in capital costs to account for replacement necessary to achieve fifty-year life of project. Interest is 4 per cent.

Long-Run and Short-Run Damage Functions

So long as improved water quality does not itself induce water-based activities, an effluent charge based upon short-run variations in external costs and provision to the waste discharger of information concerning the expected value of future charges and their duration at various levels for planning and design purposes would tend to produce optimum results. The latter would of course reflect any expected growth of water use in the region and the general level of charges would be increased over time as the external costs of a given level of waste discharge rose. However, if improved water quality itself *induces* adjustments which affect external costs, the problem of levying an appropriate short-run effluent charge becomes more complex. For example, if downstream users, say municipalities, have already installed equipment for special treatment, the short-run benefit from improved intake water quality may be appreciably less than in the longer-run because it will not be necessary to replace equipment. Or a body of water may have very little value for recreation because the water Is so poor that very few persons use it. If the quality were improved and maintained at the higher level, this would induce the development of such skills as water skiing, along with the associated investments in equipment. In these instances a certain promotional element must enter into the level of the effluent charges set, at least in the early stages of the program. As our subsequent discussion of recreation demand will show, econometric techniques promise to be able to provide us with estimates of such induced effects. But how do we take them into account in the charges approach? In these instances presuming the adjustment periods are not excessively long it is desirable to estimate a damage function incorporating longer-run adjustments and use it as the basis for charges.

If long periods of adjustment are involved the agency will then have to decide whether the present worth of the future benefits at various levels of control exceeds the present worth of the costs which would be needed to achieve it. Once again the agency would need at least approximate information on the costs of various levels of waste control and the effluent charges system would lose at least some of its information cost advantage.

Special Problems in Damage Measurement

Up to this point the costs associated with waste disposal have been broadly classified as damage costs and waste reduction costs and discussed in general terms. In this section, attention is centered on some problems of measuring damage costs. Those relating to aesthetics and recreation present particularly difficult problems of measurement. Many other costs, though they offer difficulties, are, in principle, rather straight forwardly measurable and efforts to quantify some of them are well under way.

With respect to industrial damages, attempts have been made to assess the net damages that industries suffer when they have to use low quality water. Since a variety of adjustments is possible, the problem becomes complex. Several studies have been models of ingenuity in using limited data and modelling or simulating effects on industrial operations. Some results are available for the petroleum refining, fruit and vegetable canning, thermal power, and beet sugar industries. In all these instances, industrial costs turn out to be surprisingly insensitive to intake water quality within comparatively wide ranges—especially in regard to aspects of quality that are usually influenced by prior uses and discharge of effluents. Sensitivity is greater to wastes which in most cases are of natural origin, such as chlorides and magnesium. One important reason for the comparative insensitivity is that a large proportion of water use in the heavy water-using industries is for purposes that can readily accommodate low-quality water cooling for instance. A second reason is that processes which are really sensitive to water quality high-pressure boilers and water to go in cans and bottles, for example need water of such high quality that extensive treatment is necessary for *any* kind of raw water used; water of distinctly low quality can be used with only minor incremental costs. High-pressure boiler feed water must be demineralized and the cost is fairly insensitive to the quality of intake water over a wide range. Thus, not much control of waste discharges can be justified by benefits to industrial users.

The situation is surprisingly similar for municipal water supplies. Much of what has been said about the need for high quality water supplies as a basis for preparation of potable water is more the product of emotion than of logic. The water of the much discussed Hudson River, which allegedly should not be used for municipal supply because of its poor quality, is actually of comparable quality to that at the Torresdalc intake of the City of Philadelphia and better than the quality at the intakes of the City of St. Louis. For many years acceptable drinking water has been produced from water of such quality albeit at the expense of some extra chemical and physical treatment. Moreover, a plant at Düsseldorf, Germany, withdraws water from the Rhine River, which is of far lower quality than the Delaware, the Hudson, or the Missouri, treats it with activated carbon and ozone, and produces quite suitable drinking water. Poor quality water does impose extra costs on municipal water systems but except in cases of toxic or evil-tasting substances use by municipalities ordinarily cannot justify very high levels of waste effluent control. Usually the toxic elements comprise very small and separable portions of waste effluents.

The general point of the above paragraph is brought out in a recent study of a stream system along whose shores are municipal waste dischargers and municipal water users. It involved a very careful gathering of cost data and the generation of new cost information. The study also developed a sophisticated technique for estimating by means of computer simulation the interrelationship between water quality at waste outfalls and at water intakes. It was found that vast amounts of reuse are required to justify the additional costs of advanced treatment for municipal waste disposal. For example, it turned out that the magnitude of water withdrawal for municipal water treatment and reuse downstream required to justify additional upstream waste treatment costs solely on savings to downstream water treatment plants is on the order of 16250 to 1 for small waste treatment plants and 10-30 to I for large waste treatment plants. Again it appears that the need to prepare potable water cannot justify particularly high standards of quality in watercourses.

The limited evidence from the studies and analysis discussed above leads to the virtually inescapable conclusion that higher water quality must be justified primarily on aesthetic and recreational grounds, if it is to be justified at all. This conclusion is reinforced by the results of the Delaware estuary study. This effort, which provides the basis for one of the cases we discuss in Part III, is the only full-scale regional study which has attempted to quantify the costs and gains associated with water quality improvement. The benefits of raw water quality improvement which could be identified as accruing to municipal and industrial water users were very small compared with the costs of improvement. In fact the benefits of improved water quality with respect to some parameters for some industrial purposes turned out to be negative, i.e., higher dissolved oxygen tended to increase corrosion in these uses. Let us turn then to a consideration of the aesthetic and recreational gains resulting from water quality improvement.

Aesthetics and Recreation. The adverse effects of water quality deterioration on the general environment are unfortunately obvious. Obnoxious odors of septic sewage may carry for miles; untreated municipal sewage contains floating and suspended materials that destroy the beauty of the water; and industrial effluents may render the water unattractive by coloring it. Fortunately, extreme nuisance is unlikely to occur in a stream which is used for multiple purposes and where a system of charges or other controls causes the costs of various other users to be considered in waste discharge decisions. Where it is necessary to evaluate aesthetic nuisance, however, the problem is very great.

People benefit from the aesthetic qualities of nearby bodies of water whether or not they help meet the costs of providing and maintaining water quality. There is, therefore, no incentive for an individual to express his true preferences by offering to pay for the control of the general environmental effects of waste discharges, and a market cannot develop. Not only is it impossible for the market to represent preferences for water quality control or improvement, but there is no way of parceling out and selling "goods" of this kind. For the same reasons, litigation by private parties will not adequately represent such preferences either. In the absence of usable evaluations, a judgment that extreme aesthetic nuisance is to be avoided, at least in heavily developed areas, must find almost universal consensus. Minimum standards imposed, for example, to prevent the discharge of obnoxious floating materials, may be a simple and satisfactory method of handling one aspect of this problem.

There are possibilities for evaluation, however. The aesthetic damages occurring in a river or lake in which quality deterioration does not result in extreme nuisance, are pretty well limited to those who own nearby property and those who would come from some distance to enjoy the river. Methods have been proposed and are under development to assess the value of advantages resulting to these classes of users when they are beneficiaries of recreation development. The procedure is to impute a demand curve from the enhancement of property values and the access costs of users who do not own or rent riparian lands. Such a procedure may have promise as a means of gauging recreation effects where beneficiaries can be reasonably well identified. Where a waste discharge affects both recreation and aesthetics (which except in cases of extreme nuisance may be viewed as almost joint products), a measure of this kind might yield a package value for both. Methods of estimating willingness to pay using questionnaires are also being explored. Econometric methods may be applied to these evaluation procedures to project the recreation and aesthetic values which would result from the alleviation of extremely adverse water quality conditions.

One useful study of the recreational value of water quality was under way in the Delaware estuary area at the time of this writing. Parts of the estuary experience low even zero dissolved oxygen almost every year during low-flow periods. To raise the D.O. is an extremely expensive proposition. To achieve a target of 2 ppm, the cheapest program would cost about $100 million and a similar program for 4 ppm would cost around $300 million. Imposing as these costs are, an econometric study of potential increases in participation rates shows that these higher oxygen levels might be justified on recreational grounds alone. Preliminary results suggest that placing a value on boating of about $2.50 a day, might justify maintaining 3 ppm of D.O. even if no other benefits were considered. This level serves as a surrogate for the general quality of the water. At today's levels of discretionary income, $2.50 a day does not seem a ridiculously high figure to attach to a day of boating. As already mentioned, the analysis of non-recreation benefits from water quality improvement in the Delaware, estuary suggests that they are modest when compared with the costs of improvement. The decision whether or not to achieve high levels of quality probably will turn almost entirely on recreational values. Another study of the Delaware estuary, which used quite different methods, also concluded that recreation benefits can economically justify greatly improved levels of quality in the Delaware estuary.

Another systematic study of recreational values associated with water quality was conducted by researchers at Oregon State University. This study, sponsored by the U.S. Public Health Service, focused on the Yaquina Bay, which lies on a beautiful, but comparatively undeveloped, part of the Oregon coastline. Population is small and many alternative recreational opportunities are available. Water quality in the bay was threatened by paper mill waste which might reduce its biological productivity. A relationship was envisaged between the input of angling effort and the output or yield of fish taken. Angling success per unit of effort was taken to represent the quality of the recreational experience. Demand equations based on the so-called Clawson method (which imputes benefit or willingness to pay for the recreation opportunity from travel costs) were estimated for three estuarial fisheries. Among the statistical explanations of demand was a variable representing success per unit of effort. Shifts in the demand for angling in Yaquina Bay were then obtained by applying the model to assumed reductions in the quality of the fishing resource, thus shifting the calculated demand function down. Examples of the results are as follows. If waste disposal resulted in a total loss of the sports fisheries, the annual value of this loss would be about $63,000 (consumers surplus foregone). Another way of putting the loss involved is that it would be about one million dollars if discounted at 6 per cent. This is less than the capital cost of a basic sewage treatment plant for a comparatively small city. A more realistic occurrence might be one which resulted in less than total destruction of the fishery. It would be much more common if waste disposal resulted in incremental reductions in water quality, fish life, and recreational values. Accordingly, the direct benefits of avoiding the losses associated with more normal waste disposal conditions would be even less. With a 50 per cent destruction of the fishery the consumers surplus loss would be about half as great. The authors of the Oregon State study were very careful to point out that many assumptions were involved in their analysis and that it did not incorporate the whole complex of goals and criteria appropriate for natural resources decision making.

While early results of recreation evaluation studies suggest that such analyses can be successfully done in certain instances and we can expect that in the future they will provide us with usable damage functions, routine detailed measurement of recreation damages is not now possible.

6

Collective Facilities in Water Quality Management

Introduction

A regional agency has a variety of potentially efficient collective measures to choose from: increasing streamflows during low-flow periods via releases from reservoir or groundwater storage; treatment of intake water and of municipal and industrial wastes in regional collective treatment plants in compactly developed areas; artificial or induced oxygenation of watercourses (including reservoirs and lakes); diversion of effluents via pipeline or other means; specialized stream use; artificial recharge of wastewaters to groundwater aquifers for quality improvement; and the construction of shallow oxidation reservoirs in the stream itself. These methods could supplement measures to control the generation and disposal of wastes at individual points. A regional agency seeking to minimize the over-all costs associated with waste disposal should plan and implement a system which equates the relevant incremental costs associated with water quality management in all directions. These costs include the costs of all manner of waste reduction and treatment and water supply treatment, the costs of flow regulation (including all opportunity costs that arise from the value of stored water for alternative uses), as well as the costs of conforming waste discharge to streamflow, changing industrial processes and outputs, and changing industrial location, and, finally, any damages which result from residual water quality deterioration.

This statement that the agency should equate the "relevant" marginal costs does not mean that the marginal costs of waste reduction should be equal at all points of waste discharge. The costs of treatment or other waste reduction measures will tend to be higher at upstream points because upstream activities are reflected in downstream quality and costs. What the rule does mean is that it must not be possible to make marginal "trade-offs" and thereby reduce damage and waste reduction costs.

Ideally, the results of all relevant water resources systems and operating procedures would have to be considered and a solution derived which simultaneously indicated the optimum combination of systems elements and operating procedures. In industrial areas the solution would entail a system of charges and/or other measures such as effluent standards and zoning designed to secure the optimum amounts and locations of waste discharges. An "efficient" solution would enable the maximum net benefit to be obtained from the available water resources, and, as the latter implies, the over-all cost associated with the disposal of an optimum amount of wastes would be minimized.

In economic terms "cost" means the value of foregone opportunities. In regard to water quality management such opportunities may be associated with resources used, for example, to build and operate treatment plants or reservoirs which could be used productively in other applications, or they may consist of opportunities to use water for purposes other than waste carriage or assimilation. The development in this chapter is simpler to visualize if the only output of the system is water quality control, so that all costs associated with waste disposal consist either of resources devoted to control facilities (reservoir regulation of streamflows, treatment of wastes, process changes, effluent diversion facilities, etc.) or of damages associated with water quality deterioration. The principles are exactly the same, however, if some of the costs consist of foregone benefits from other water-related outputs such as peak power.

Basin-Wide Firm Again

The basin-wide firm is used here to illuminate economic principles. It is not put forth as a policy recommendation, for it would, in reality, have grave disadvantages: It cannot take account of uses that are not reflected in market prices; and, as a monopoly, it would present problems of regulation.

To illustrate some of the principles involved in taking account of the technical interdependencies in a river basin when economies of scale in water quality management measures are available, let us assume that a single firm (1) conducts all water-using industrial enterprises, all water and waste treatment facilities (no privately owned septic tanks, water softners, etc.), and all water transportation and related facilities; (2) operates all hydroelectric facilities, owns all land and structures in the flood plain, and is the sole provider of flow regulation; (3) controls all recreational uses of the watercourse; and (4) operates in competitive markets or in ones where public regulatory authorities set prices equal to marginal costs at levels of output that just clear the market.

In order to maximize its profits the firm would select the combination of water quality control measures (water supply treatment, waste generation and treatment, flow augmentation, wastewater releases coordinated with streamflow, etc.) and downstream damages that would minimize the over-all costs associated with its waste disposal at its most profitable level of activity. This would be accomplished by equating the relevant incremental costs associated with waste disposal for all alternatives. As previously indicated, these costs would include any valuable outputs that are sacrificed to improve the efficiency of the waste disposal system.

If collective treatment costs are less than treatment at individual discharge points, the firm would transport effluent to collective treatment plants until the marginal costs of transportation and marginal treatment saved were equal. Since transporting wastes long distances and building scattered treatment plants are usually expensive procedures, the firm would weigh these and other diseconomies of scattered development against the economies (less congestion) to arrive at the optimum compactness of economic activities.

Compact economic development might be influenced by certain physical characteristics of the stream. Since the rate of reaeration through the air-water interface is directly proportional to the oxygen deficit, the organic waste degradation capacity is a decreasing function of the dissolved oxygen quality standards. This, plus the possibility of stream treatment to take advantage of economies of scale, emphasizes the possibilities of stream specialization (i.e., using some stream or stretches of stream more heavily for waste disposal and maintaining higher quality in others) as a means of reducing the costs associated with waste disposal. Again the firm would trade off such possibilities against alternatives such as additional flow augmentation, higher levels of treatment, and other means of waste reduction. In deciding upon the location of a new industrial plant, the firm would consider not only transportation costs and other factors affecting assembly, production, and distribution costs but also the increment of total waste reduction and damage costs of alternative locations. The location producing lowest costs would be the one at which the marginal costs of the alternatives bearing upon over-all costs are equalized.

While the point is expanded later, it is worth noting here that the firm "internalizes" all these costs. A water resources agency in a basin cannot internalize all of them, and *consequently it must attempt to have the opportunity costs of water resources use, including waste disposal, reflected in the decisions of entities that are simultaneously considering other resources costs.* This can be done by standards or, preferably, by charges.

The firm's general objective is to maximize profits, which implies that the costs associated with waste disposal will be minimized at the profitmaximizing level of output. If, for example, the firm could lower its costs by doing a little less waste treatment and permitting a little more damage, or by doing a little more water treatment and a little less waste reduction, or by a little more augmentation of low flows and a little less temporary storage of wastes, or a bit more process adjustment and a little less direct reaeration of the stream, etc., its over-all profit position could not be at a maximum. As part of its general profit-maximizing activities, the firm would integrate its waste control activities with other aspects of its water-related operations. One major linkage is through flow regulation.

Thus, in computing the costs of alternative water quality control devices, the firm would need to consider complementary and competitive relationships among different water uses and alteration of the stream's flow regimen. Accordingly flow-augmentation costs would have to be determined in light of the fact that this alternative is often complementary with navigation, for example, but at least partly competitive with irrigation, flood damage reduction, power, and recreation. Any net benefits associated with other uses foregone in using flow augmentation to improve water quality in the optimum system are counted among the costs associated with water quality improvement. Moreover, the firm would consider the full marginal costs of producing particular products including the costs imposed on other activities by waste disposal. If the market adequately registered the population's evaluation of all goods and services sold by the hypothetical firm, its solution to the waste disposal problem would be "efficient," as that term is understood in economic welfare theory.

Many extreme assumptions have been made in developing this example; hence the basin-wide firm is not put forward as a realistic policy alternative. Still, the concept has provided a useful vehicle for illustrating applicable systems and decision criteria. The following points concerning public policy merit particular emphasis.

First, public policies which explicitly recognize water quality management as a problem of optimizing an interdependent system in a region will find many alternative control measures to consider. Accordingly, a least-cost system is more likely to result from the implementation of such policies than from the adoption of narrowly circumscribed conventional approaches. The case studies in the next chapter show the savings that can result from a more comprehensive and flexible approach.

Second, implementing the regional approach requires the planning, design, and integrated operation of many interdependent elements in the system. The illustrative firm "internalized" all these elements and, under our assumptions, found their optimum planning and operation in its interest as a matter of seeking maximum profit.

A water management agency should not attempt to internalize more elements than needed for its objective. In one instance, an agency may plan, design, and operate only collective measures such as reservoirs and reaeration devices and use direct controls or incentives (charges) to integrate these with the rest of the relevant economic system. In other instances a regional agency may find that it serves efficiency to go much further and it may plan, design, and operate virtually all water-related facilities in a region including water and wastewater treatment plants. Even in this case it would still be necessary to influence industrial process design and plant location. This could be done by direct regulation or, again we think preferably, by an appropriate system of charges.

In the following section we develop these thoughts further by using a somewhat simplified example which nevertheless permits us to address some matters of planning incentives and financing in a system which includes collective measures.

Regional System with Flow Regulation

We assume a system where the economically relevant alternatives are low-flow augmentation and waste reduction activities at individual points of waste discharge. The regional agency will design and operate a reservoir for flow regulation and use a system of effluent charges to control waste discharges at outfalls. If the reservoir is multipurpose we assume that the agency takes appropriate account of complementary and competitive uses of the water in determining the marginal cost of flow regulation for water quality management. For example, if navigation benefits occur in a fully complementary fashion with water quality benefits, the two must be summed and compared with the marginal costs of flow regulation.

Costs in this case are the total costs at various levels of flow of optimal combinations of downstream alternatives-for example, waste treatment, water supply treatment, and value of residual physical damages. We assume that these combinations are induced by an optimal system of effluent charges in accordance with the analysis.

The slopes of the total storage cost and downstream cost functions are equal to their respective incremental functions. Accordingly at point *X* the optimality condition (i.e., equality of incremental costs) prevails. The same solution can readily be stated in terms of functions which are more familiar from previous discussion. In general the optimality criterion is that the *expected value* of the incremental costs of all relevant alternatives should be equalized at the margin. If effluent charges are used to control individual waste discharges, the optimal condition requires that such charges reflect the incremental offsite costs imposed by each waste discharge. Charges will of course be lower when the assimilative capacity is increased by flow augmentation and/or by stream reaeration. In a correctly planned system the lowered charges reflect a gain in efficiency.

It may be worthwhile restating why point X represents an optimum combination of upstream and downstream measures. At this point a small further increase in reduction activity at the outfall would cost more than the expected value of downstream water treatment costs and residual damages avoided by downstream users. The argument is symmetrical for a slight decrease in reduction activity at the outfall.

When costs associated with waste disposal are minimized, the incremental costs associated with waste disposal are equated at each outfall. This means that the marginal costs associated with waste disposal are equalized in all directions. However, it does not mean that the marginal cost of reduction will be the same at each outfall along a stream. This would only be the case if downstream costs per unit of waste were equal for different outfalls, as they might be when two outfalls are across the stream from each other; then strict marginal cost equality would be necessary for cost minimization.

Effluent Charges and Collective Measures

Fortunately, devising an appropriate effluent charges policy for an agency which incorporates economically desirable collective measures in its system requires no new principles whatsoever. The ideal charge will still be equal to the marginal external cost imposed. The residual external marginal cost imposed by waste discharge, when control at relevant points of waste discharge is induced to an optimal level by effluent charges, is the measure of the marginal benefit of the collective quality improvement facility—say a flow-regulating reservoir.

Let us assume (unrealistically) that the agency creates the entire assimilative capacity of the river by producing a flow of water.

Assume further that this flow can be produced at a constant marginal cost. Finally, assume that assimilative capacity can be increased efficiently in small increments. Under these assumptions, the agency, if acting efficiently, would increase assimilative capacity in such a way that the marginal cost of the last increment in capacity to control the effect of a waste would be equal to the effluent charge on that waste, which in turn is equal to the marginal cost of waste control at outfalls our "equal marginal cost" principle again. In this instance the proceeds from the effluent charge reflecting the marginal external cost would just cover the total cost of producing the assimilative capacity.

Given the same situation, except that we now assume that the marginal cost of further increments of capacity is rising, the relationship between the yield from charges and the total cost of the investment will change. The yield from effluent charges will now exceed the amount necessary to pay the total cost of the investment when optimum capacity is attained. Conversely, if the marginal cost of expanded assimilative capacity declines over the relevant range, the proceeds from the charge will be insufficient to cover the cost of the capacity expansion. This will be true even though the capacity expansion reduces the over-all costs associated with the regional waste disposal system. If the regional agency decides not to augment assimilative capacity but to reduce waste loads by measures such as collective treatment, it should still follow the equal marginal cost rule and charge waste disposers the marginal reduction cost (which equals the marginal external cost) of handling the particular kind of waste involved.

The assumption that the regional authority created all of the assimilative capacity of the stream makes for a theoretically neat outcome, but it is unrealistic because a stream always has some natural waste assimilative capacity. This circumstance in no way affects the use of the marginal cost rule as the appropriate criterion for the agency's investment decisions, but it does affect the financial results. When the agency levies a charge reflecting marginal external costs in the presence of natural assimilative capacity, part of the payment made is a "pure rent" on the scarcity value of the natural capacity. As we have indicated earlier, this is analogous to a charge levied on the use of public lands for grazing, for example. Accordingly, when natural assimilative capacity exists, the proceeds from the charge will be more than sufficient to pay for an economically justified augmentation of assimilative capacity produced under constant cost conditions. Any excess revenue should, in accordance with our discussion in Chapter 6, be used for general public purposes. If an agency implemented only economically justifiable facilities and levied charges equal to the marginal external cost, it might take rather sharply decreasing costs in the provision of collective facilities for the agency to incur a deficit if there is a large amount of naturally occurring assimilative capacity initially available.

The above discussion has assumed that collective facilities can be expanded more or less continuously. Since some of them, such as reservoirs, involve substantial indivisibilities, this is not very realistic for small systems. But the principles are the same even when durable and lumpy investments are made. Of course it will be necessary to make long-run projections of population and industrial production—and their related waste discharges—with their attendant uncertainties. Other measures such as mechanical reaeration can be expanded more or less continuously. Groundwater recharge operations represent an intermediate situation. Moreover, as a system becomes larger and more diverse, even absolutely large additions become small relative to the total system and can be implemented on the basis of comparatively short-run considerations.

The economic principles which we have discussed in this section are developed in a rigorous mathematical form in the appendix. Indeed, the appendix provides a compact synthesis of most of the economic principles which we develop in this book.

Unmeasured Values as Constraints on the Objective

If there were no problems in the manner in which the market registers water quality values, the policies described above would produce a solution quite consistent with the general rationale of a market system. But the market provides no value, or no satisfactory value, for some aspects of water quality, and this creates a special problem for public bodies because they must take account of all aspects. Considerable progress has been made in finding ways to assess the worth of such matters as general aesthetic effects, public health, and recreation, but dealing with these unmeasured values is still far from being a routine matter.

Two ways of handling unmeasured values come to mind. One way would be to label them "intangible" and disregard them in planning and designing water quality management systems. Then, when the proposed systems are presented for consideration and authorization by representatives of the public, side information could be provided on aesthetic effects, public health, and other matters considered relevant to arriving at a decision in the public interest.

The second approach is to include hypotheses about such values in the process of system planning by expressing them in physical terms and treating them as constraints upon the cost-minimization objective. For example, if social choice should dictate that the oxygen level in a stream must be high enough to support fish life at all times, the system must be planned to minimize the real cost associated with waste disposal subject to the constraint, or "minimum D.O. standard." This may require a different combination of units with different operating procedures than a system planned without the standard or constraint. If the constraint is effective and not automatically achieved when measurable costs (including damage costs) are minimized, it will increase the cost of the system, the extra cost representing the limitation which the constraint places upon the objective.

To be consistent with efforts to achieve maximum welfare, the planning process should consider constraints "provisional" and view them as matters for research and study in order to discover how well they represent the preferences of society. One way of studying them from this point of view is to test their cost sensitivity. Varying a constraint by small amounts, redetermining the optimum system, and relating the change in costs to the associated physical changes in specific stretches of stream will provide information that will permit considered choices to be made by political representatives.

One useful way of stating the results of experiments with the constraints that are not valued directly by, or imputable from, the market is in terms of what they must "at least be worth." For example, a social judgment may be made that dissolved oxygen is to be maintained beyond the point indicated by the cost-minimizing solution in order to preserve or enhance a recreational fishery. A comparison of the system with and without the constraint will not establish precisely what preserving the fishing pleasure is worth, but it will indicate the *least* value that must be attached to it if the higher level of control procedures is to be worthwhile. Or the additional cost of a higher level of certainty of achieving the desired dissolved oxygen level can be estimated. If constraints are imposed representing goals not directly commensurable with the values stated in monetary terms in the objective function, marginal conditions analogous to those indicated earlier must still hold if the cost-minimization objective is to be fulfilled. The optimum system is not attained until a situation is reached in which it is impossible to make incremental "tradeoffs" between alternatives that will lower costs without violating the constraint.

Once a combination of system elements has been decided upon, it will be necessary to utilize charges, effluent standards, or some other regulatory device to achieve optimum waste reduction at individual outfalls. The charge, if a charge is used, must be just high enough to achieve the standard, and it must not be possible to expand the largescale measures at an incremental cost less than the incremental cost which can be avoided at the individual outfalls. If several wastes are involved, the expected value of all incremental costs avoided by an expansion in the large-scale measures must be added together and compared with the cost of expanding the measure.

We may generalize by saying that once a constraint is established by a duly constituted, well-informed political decision-making process, it may be viewed as representing a damage function which is perfectly inelastic at the specified quality level and degree of certainty. When this view is taken, the analysis with respect to appropriate levels of effluent charges and investment criteria presented in earlier sections and in the chapter appendix holds exactly.

Large amounts of data and extensive knowledge of physical and economic relationships are needed to implement the procedure outlined above. Rapid and flexible computational techniques for estimating the physical characteristics and patterns of water quality and for carrying out the actual minimization (maximization) procedure are also essential. Some of these techniques, as well as the specific character of the information required to make reasonably dependable analyses of complex ri basin systems, including cost sensitivity analysis of constraints.

Here we simply note that once a system is implemented the charges themselves are important sources of information about investment decisions in collective facilities. They indicate the marginal cost which waste disposers are incurring in order to meet the standard. This information can be used by the basin agency to determine whether a collective measure, either by capacity expansion or improved operation, can maintain the standard at a lesser incremental cost. Thus once the system is established the agency can obtain information on costs incurred by waste disposers without having detailed knowledge of their cost functions.

Institutional Restraints. Where there is no regional agency with authority to plan and operate a comprehensive system of quality management, the type of cost-sensitivity analysis outlined above can be used to illuminate the desirability of institutional change. In most basins in the United States the only collective measures are industrial and municipal waste-treatment facilities and reservoirs operated for low-flow augmentation by federal or state agencies. By comparing a least-cost system restricted to these measures with one that incorporates all relevant alternatives, a price tag can be put on the restriction. Or, to put it another way, the comparison will provide an estimate of the economic gain that society can derive from an institutional mechanism with authority to implement a wider range of alternatives. This is illustrated by the Potomac case reported in the next chapter.

Some "Short-Run" Aspects of System Operation and Cost Assessment

At this point it is useful to introduce once more the distinction between "long-run" or "planning" costs and "short-run" or "operating" costs. There are two reasons for making this distinction in the present context: First, a system should be operated on the basis of its "current" opportunity costs, especially when certain elements such as dams must be introduced in "chunks." Second, hydrologic or meteorologic fluctuation may make it desirable, in terms of cost minimization, to operate the water quality control facilities in a variable manner. This indicates that, aside from any informational and administrative costs, effluent charges should be varied so as to maintain equality in the short-run marginal costs of alternatives.

So far the discussion has been primarily in terms of planning the regional system, and it was assumed that choices were not limited by past decisions. The view taken was sufficiently long-run, for example, to permit the building of dams *instead* of treatment plants. For planning, these are valid and appropriate concepts.

The point has been made several times that if the objective is to minimize the value of resources used internally and externally in connection with water quality management, the relevant cost for choosing a particular alternative is the value of the opportunities that must be foregone if it is chosen. This concept of "opportunity" costs has been used throughout, and methods of producing a better reflection of the value of foregone opportunities to society in decisions with respect to waste disposal.

The *concept* of opportunity costs does not change when a particular set of facilities has been decided upon and installed, but the *character* of the opportunities foregone does. When a dam is in the planning stage, all the labor, cement, steel, etc., needed for its construction can still be used in other activities, but once the labor, cement, etc., have been embodied in a dam they can no longer be used in alternative activities, and their price can no longer be considered an opportunity cost. The only costs that are still opportunity costs and thus relevant for further decisions are those that arise when a dam is operated for a particular purpose. If the dam is multipurpose, the opportunity costs may be largely "internal," representing the net value of the other uses foregone when its capacity is used for a particular purpose, although some operating costs will be involved, too.

In planning water quality management systems, therefore, the definition of costs should include all capital costs of new facilities. The system for water quality improvement should be expanded to the point where the cost of reducing an *additional unit* of any given waste or combination of wastes if the reduction activity involves joint products raises total reduction costs as much as it diminishes water quality damages. If there is a constraint, the system should be expanded until the constraint is met and the marginal total costs of all alternative measures for achieving the constraint are equalized.

However, after a water quality management system is established, only the currently variable costs and opportunity costs internal to a multipurpose system are relevant. For example, assume that a decision is to be made about whether to retain a waste treatment plant with excess capacity or replace it with a smaller plant. Comparison of total costs, including capital costs of both plants, would almost certainly lead to the wrong conclusion. If the original plant has no alternative use, the relevant costs include only the operating costs, the land, and portions of the existing investment for which there are alternatives uses. The value of such uses must, of course, be brought into the comparison because they represent genuine opportunity costs. The test is always whether there are currently feasible, valuable alternatives. From the point of view of allocating resources to their most productive use, only inputs which have alternative uses can be said to involve costs.

Since storage reservoirs ordinarily cannot be economically constructed in such a way as to add small increments to system capacity, a system adjusting to changing waste load conditions will have periods of excess capacity. Under these circumstances the short-run opportunity costs of using the reservoir for flow augmentation for water quality improvement must be compared with the opportunity costs of alternative uses of the reservoir capacity. Appropriate operating procedures of the system would be based upon relevant marginal costs. Reliance upon flow augmentation would be heavier in the early period after the dam was built, and effluent charges or standards would be accordingly lower in those years.

This means that care must be taken to provide waste disposers with information on longer-run prospects so that investment decisions will not be based on the presumption that charges will continue to be low.

Summary

This chapter has focused on the integrated planning of various types of waste load reduction, flow regulation and in-stream measures; the incorporation of "unmeasured values"; and some "short-run" economic problems which stem from environmental variability. Attention was directed to the important role that can be played by effluent charges-either alone or in conjunction with other measures for regulating waste discharges – in articulating the decisions of individual waste dischargers efficiently into a regional system for water quality management. In addition, it was pointed out that the assessment of charges is important in financing the collective features of such a system.

It was indicated that, ideally, system planning would consider all alternative uses of water, including its use for waste disposal, the effects of the various uses on water quality, the losses imposed on other uses by quality deterioration, and the value of water-derivative uses. All feasible system plans and operating procedures would be considered, and a solution would be derived which would indicate the optimum combination of system elements and operating procedures in light of the objectives and constraints relevant to the system. The regional system view, taking cognizance of the efficiency of collective measures, did not require the introduction of any new economic principles. But there are practical reasons, including the inherent complexity of the problem and the primitive state or absence of much of the required data, why an optimum solution cannot be fully attained. Furthermore, some inevitable arbitrariness in the specification of constraints, the absence of fully satisfactory means of handling the risk inherent in natural phenomena, and the uncertainty involved in forecasting future economic variables, militate against the achievement of fully optimum decisions.

Nevertheless, an approach of the general type will improve decisions on water quality management. It emphasizes the importance of identifying interdependencies and viewing areas and economic functions tied together by "spillovers" in a "system" context; requires an explicit and unambiguous statement of objectives and constraints; stresses the identification and quantification of alternatives and their systematic testing in light of criteria of merit developed in terms of the objective or objectives of the system; and endeavors to cause the opportunity costs of water resources use to be reflected in decisions not controlled directly by the water resources management agency. These features of the analysis represent a way of thinking about the problem of water quality management; they do not imply a specific procedure for solving it. When the situation is a complex one involving numerous alternatives and constraints, optimization procedures applied to formal mathematical models, often with the aid of computers, can be very helpful.

❑❑❑

7

Benefits of Groundwater Resources

Definition and Meaning

Groundwater may be defined as subsurface water that occurs beneath the water table in soils and geologic forms that are fully saturated. It is an integral part of the hydrologic cycle, and any approach to groundwater problems should recognize this. Groundwater is not only an important natural resource but also an essential part of the natural environment. Much of the folklore and many of the widely held misconceptions about the nature of groundwater for example, that it occurs in underground lakes, rivers, and veins and can be detected by listening for the noise are dispelled by Lehr and Pettyjohn in their description of groundwater and its flow patterns.

The global hydrologic cycle consists of the movement of water between the oceans and other surface water, the atmosphere, and the land. Only 2.7% of all the water on the planet is fresh, and of that only 0.36% is easily available to users. It has been estimated that about 40,000 billion gallons per day (bgd) pass over the conterminous United States as water vapor. Approximately 10% of this, 4,200 bgd, is precipitated as rain, snow, sleet, or hail, equivalent to an average uniform annual rainfall of 30 inches nationwide. About two-thirds of the precipitation evaporates or is transpired by vegetation. The remaining 1,450 bgd accumulates in ground and surface waters, flows to the sea or across national boundaries, is consumed, or evaporates from reservoirs.

Table 7.1. Estimate of the Water Balance of the World

Parameter	Surface Ar. $(km^2)x10^6$	Volume $(km^3)x10^6$	Volume (%)	Equivalent Depth(m)	Residence Time
Oceans and Seas	361	1370	94	2500	~4000 yrs
Lakes and Reservoirs	1.55	0.13	<0.01	0.25	~10 yrs
Swamps	<0.1	<0.01	<0.01	0.007	1-10 yrs
River Channels	<0.1	<0.01	<0.01	0.003	~2 wks
Soil Moisture	130	0.07	<0.01	0.13	2 weeks-1 yr
Ground water	130	60	4	120	2 weeks-10,000 yrs
Icecaps and Glaciers	17.8	30	2	60	10-1000 yrs
Atmospheric Water		504	0.01	<0.01	0.025 ~10 days
Biospheric Water	<0.1	<0.01	<0.01	0.001	~1 wk

Of the 1,450 bgd only 675 bgd are usually available for intensive beneficial uses. The ability of an aquifer to store and transmit water is a function of its permeability and porosity. When the saturated substratum is sufficiently permeable to store and transmit significant quantities of water, the geological formation is called an aquifer. There are two main types of aquifers, confined and unconfined. An unconfined or water-table aquifer contains water under atmospheric pressure; the upper 19 surface of the water is called the water table and may rise and fall according to the volume of water stored, which is dependent upon seasonal cycles of natural recharge. A perched aquifer is one in which a limited layer, or lens, of impermeable material occurs above the water table, forming a thin zone of saturation above it. It is a type of water-table aquifer. The second major category of aquifer is the confined, or artesian, aquifer. These are bounded top and bottom by layers of relatively impermeable geologic formations termed aquitards, or confining layers. The aquifer is completely saturated with water that is under greater than atmospheric pressure. An artesian aquifer is not recharged everywhere uniformly, but in one or more general recharge areas. Water levels in non-pumping wells of unconfined aquifers correspond to the level of the water table and therefore vary according to the volume of water in storage. Water levels in nonpumping wells tapping confined aquifers are dependent upon the artesian pressure of the water in that aquifer, and in some cases the water may exceed the top of the well casing, thus causing a flowing well. The hypothetical projection of such water levels is known as the potentiometric surface. The aquitards, or confining layers, are not totally impermeable and permit some recharge or discharge to lower confined aquifers. Aquifers can occur in unconsolidated materials such as sand and gravel or in consolidated material or bedrock. The latter may consist of carbonate rock, volcanic rock, or fractured igneous, metamorphic, and sedimentary rocks. Sand and gravel aquifers usually contain the most groundwater, but high-yield wells can also occur in carbonate and volcanic rocks.

Groundwater Movement

Groundwater moves in response to gravity, pressure, and friction, the first two driving the water, the latter resisting motion. Due to the complexity of the channels through which groundwater flows, it is difficult to construct a model of the movement of groundwater at a microscopic level. The French engineer Darcy, however, formulated an empirical law in the mid-nineteenth century that effectively averages the microscopic complexities, providing a macroscopic model for groundwater movement. Darcy's Law relates the rate at which groundwater flows across a surface to the rate of change of energy of the groundwater along the flow path. Under ideal 20 homogeneous isotropic geologic groundwater conditions, the average linear groundwater velocity ((v)) can be expressed by the Darcy relation:

$$v = - [K\, dh/dl]/n$$

where: v = the average linear groundwater velocity
K = hydraulic conductivity
dh/dl = hydraulic gradient
n = transport porosity

For both granular and fractured media, the same equation is used to compute values for average linear groundwater velocity. The great difference is in the value of *n*. For granular media n nearly represents the total porosity; in fractured rock or clay n represents the total void space in connected fractures within a unit volume of media. The value of n in granular material is usually between 0.3 and 0.45; in fractured material n is usually very small, of the order of 10*2 or 10*4. When the values of *K* and *dh/dl* are similar in granular and fractured media, the average linear groundwater velocity may be orders of magnitude larger in fractured media than in granular media. The velocity of groundwater in a horizontal sand and gravel aquifer is usually in the range of 0.05 to 1 meter per day.

Darcy's Law is central to the derivation of equations used to model the flow of groundwater. Mathematical models, based on the physics of groundwater flow and boundary conditions imposed by the groundwater basin in question, usually take the form of a boundary-value problem. Bear discusses the application of the three methods used for solving such boundary-value problems, namely analytical, analog, and numerical. Analytical methods result in explicit mathematical expressions of the solution and are useful for problems where the governing equations may be simplified and the boundaries of the basins idealized. Analog models are scaled physical models usually involving the construction of electronic equipment to simulate groundwater flow. Numerical mathematical techniques are the basis for digital computer simulation of transient groundwater flow in aquifers. Digital simulation of groundwater flow requires an expertise in computer programming and its use is bound to digital computers. However, it is more flexible in its ability to handle irregular groundwater boundaries and variations of groundwater movement through time and space than are the other two modeling techniques. The two most widely used numerical methods for solving groundwater equations are the finite-difference and finite-element techniques. The main difference between groundwater and surface water is that the movement of groundwater takes place very slowly. Mixing of groundwater is slow, in contrast to the good mixing potential in the turbulent flow of most surface water. This factor is extremely important when considering the fate of contaminants.

Occurrence and Natural Quality of Groundwater

Most aquifers occur within 2,500 feet of the surface of the land. They may be thick or thin, extensive or local, very near the land surface or at considerable depths. It has been estimated that 30% of the stream flow of the United States is supplied by groundwater that emerges as natural springs or other seepage. In certain areas in times of drought, the stream flow in low-flow months may be largely provided by groundwater. The interrelatedness of surface water and groundwater is further underlined by the fact that under certain conditions seepage from lakes, rivers, streams, reservoirs, and canals may recharge aquifers.

The quantity of groundwater in storage is much greater than the volume of surface water available in streams and lakes. It comprises more than 96% of all the fresh water in the United States, the remaining 4% occurring in lakes, rivers, and streams. Estimates of groundwater within 2,500 feet of the land surface in the conterminous United States range from 33 quadrillion to 59 quadrillion gallons to as much as 100 quadrillion gallons. As a point of comparison, Lake Michigan contains 1.3 quadrillion gallons of water. Not all of the groundwater in storage is available for use, however, because some is bound to soil particles. The cost of extraction may also restrict its use in certain areas. The country has been divided up into 10 groundwater regions. This is considered the best broad classification of the groundwater situation in the conterminous United States. The division is based on the types of aquifers.

1. The western-mountain ranges consist mainly of igneous, metamorphic, and consolidated sedimentary rocks. Most of the groundwater occurs in rock fractures. The large amount of precipitation that falls in this area recharges aquifers in adjacent regions.
2. Alluvial basins consist of valleys surrounded by mountains in the arid southwest. Water levels in many aquifers have declined over recent years due to the use of large volumes of groundwater for agricultural irrigation.
3. The Columbia Lava Plateau consists of lava flows and unconsolidated sediments. In some areas there are large supplies of groundwater.

4. The Colorado Plateau has high, dry plateaus of sedimentary shale and sandstone and has a scarcity of productive aquifers.
5. The High Plains are extensive and semi-arid to sub-humid. The unconsolidated sedimentary rocks form the Ogallala Aquifer. Due to extensive agricultural irrigation, withdrawals have greatly exceeded groundwater recharge in some areas.
6. The glaciated central lowlands consist of glacial drift, sand, and gravel forming major groundwater reservoirs. In some heavily populated or highly industrialized areas the groundwater has been overdeveloped or polluted.
7. The unglaciated Central Region is composed of horizontal consolidated sedimentary rocks, with limestone and sandstone formations providing the major aquifers. The yields are low to moderate and adequate for domestic supply but would not be sufficient for irrigated agriculture.
8. The unglaciated Appalachians have high-yielding limestone and sandstone aquifers, and lower yielding shale aquifers. The water table may occur at a considerable depth below the surface. This region has an abundance of surface water year round.
9. The glaciated Appalachians region is similar to the unglaciated Appalachians.
10. The Atlantic and Gulf Coast Plain has an abundance of groundwater and surface water. The geology consists of unconsolidated gravel, sand, silt and clay, and limestone. Supplies of groundwater may not be abundant in some areas of Texas, Louisiana, Mississippi, and Alabama. Florida has a prolific aquifer.

Unfortunately there is more than one classification system for dividing the country up into water regions. The system used by the U.S. Water Resources Council is shown. For the purpose of compiling and analyzing water-resources data for both surface water and groundwater, the U.S. Water Resources Council divided the nation into 21 major water-resources regions, 18 within the conterminous United States, and the other 3 being Alaska, Hawaii, and the Caribbean area. These 21 regions are further subdivided into 106 subregions. The regional divisions are areas that contain either the drainage area of a major river or the combined drainage of a series of rivers.

The EPA used another classification system in compiling data on groundwater pollution for various regions of the United States. The EPA system most closely resembles that described by Lehr, which is summarized above.

Along with the geological characteristics of the groundwater regions, climatic conditions in particular, precipitation and temperature have effects upon the quantity and quality of the groundwater available. Variations from average precipitation may cause droughts or floods. Water is lost from the land to the atmosphere by evaporation and transpiration, or evapotranspiration. When evapotranspiration exceeds precipitation, groundwater recharge by percolation does not occur. In the arid areas of the Southwest, for example, annual precipitation is less than 10 inches, and potential evaporation could exceed this 4 to 20 times. In such areas groundwater recharge occurs mainly in wet, or multi-year, cycles. In the eastern states, the annual precipitation exceeds the evaporation, thus providing surpluses that contribute to stream flow. Regional variations in temperature may also affect groundwater in ways unrelated to surface evaporation. For example, frozen ground does not permit infiltration of rain, thus causing floods and preventing groundwater recharge.

The United States Geological Survey (USGS) published summary appraisals of the nation's groundwater resources between 1984 and 1992 for each of the regions that were briefly outlined above in the U.S. Water Resources Council scheme. A listing of the titles is given. These geological survey professional papers list groundwater usage for each region by categories of use (e.g., municipal water supply and agricultural irrigation). They list the natural groundwater quality and any problems that may be encountered due to the presence of natural iron in the water or to its salt content and hardness. In addition, they list present and potential anthropogenic threats to the quality of groundwater.

The National Water Summary also put together by the USGS, contains descriptions of the occurrence, use, and general quality of groundwater resources of each state. Each summary contains: (1) the physiographic, hydrologic, and geologic framework of the state's groundwater system; (2) a description of the principal aquifers; (3) groundwater withdrawals and water-level trends; (4) a description of the state's groundwater management program(s); (5) maps showing geographic distribution of the principal aquifers; (6) figures illustrating the areal distribution of major groundwater withdrawals and trends in water levels; and (7) selected references.

The quality of groundwater is often described in terms of hardness and salinity. Hardness reflects its calcium and magnesium content and is usually expressed as the equivalent amount of calcium carbonate. It can be viewed as a measure of usefulness for domestic and industrial purposes.

Water quality, however, is usually defined in terms of the concentration of its chemical constituents. As water moves through the hydrologic cycle, it interacts with the atmosphere, soils, and subsurface geologic formations, all of which affect its chemical composition. Thus, there is a natural background level for the chemical content of the water, and this level varies regionally and may be subsequently augmented from industrial and domestic sources. Freeze and Cherry discuss the natural chemical content of groundwater in detail. Rainwater is saturated with oxygen, nitrogen, and carbon dioxide gases and is usually slightly acidic, having a pH of about 5.6. The acidity may be increased by industrial pollutants, namely, the oxides of sulfur and nitrogen. The more acid the rainwater, the more likely it is to react with the geologic materials with which it comes into contact.

Table 7.2. Simple Groundwater Classification Based on Total Dissolved Solids

Category	Total Dissolved Solids (mg/l or g/m^3)
Fresh water	0-1000
Brackish water	100-10,000
Saline water	10,000-100,000
Brine water	More than 100,000

Rainwater percolating through the soil may increase in acidity due to the biological processes that occur in that zone. Plant and microbial respiration produce carbon dioxide, which increases acidity. It would be possible for the percolating water to become supersaturated with carbon dioxide. Acidity may also be increased by products of decomposition such as humic and fulvic acids, nutrient uptake by roots, and nitrifying bacteria, but these are minor factors compared with the amount of carbon dioxide produced by respiration. Thus, the water undergoes a chemical change during its passage through the soil to the underlying water-bearing formations.

Table 7.3. Classification Based on Dissolved Inorganic Constituents in Groundwater

Major Constituents (greater than 5 mg/l)	
Bicarbonate	Silicon
Calcium	Sodium
Chloride	Sulfate
Magnesium	Carbonic acid
Minor Constituents (0.01-10.0 mg/l)	
Boron	Nitrate
Carbonate	Potassium
Fluoride	Strontium
Iron	
Trace Constituents (less than 0.1 mg/l)	
Aluminum	Molybdenum
Antimony	Nickel
Arsenic	Niobium
Barium	Phosphate
Beryllium	Platinum
Bismuth	Radium
Bromide	Rubidium
Cadmium	Ruthenium
Cerium	Scandium
Cesium	Selenium
Chromium	Silver
Cobalt	Thallium
Copper	Thorium
Gallium	Tin
Germanium	Titanium
Gold	Tungsten
Indium	Uranium
Iodide	Vanadium
Lanthanum	Ytterbium
Lead	Yttrium
Lithium	Zinc
Manganese	Zirconium

In the saturated zone, the water reacts with the geologic formations, increasing the content of total dissolved solids. Thus, the chemical quality of groundwater depends both upon its age and the geological formations encountered in its flow history. Carbonate formations would increase the magnesium and calcium content and also the concentration of bicarbonate due to the dissolution of calcite ($CaCO_3$) and dolomite (MgòCaCO_3). Aluminosilicate minerals would increase the concentrations of sodium, potassium, magnesium, calcium, and silicon hydroxide ($Si(OH)_4$) in the water.

In consolidated deposits consisting of minerals from a range of sedimentary, igneous, and metamorphic sources, the order of encounter of the groundwater with the different assemblages determines the chemical constituency at a particular point in time. Sulfate-bearing minerals such as gypsum and anhydrite, although they occur less frequently than the carbonate and crystalline formations, are characterized by high solubilities. Thus, in older groundwater that has encountered sulfur-bearing minerals, the sulfate anion may dominate the bicarbonate anion. In very deep, old groundwater, the chloride anion may dominate both the bicarbonate and sulfate ones due to the presence of readily soluble minerals such as halite (NaCl) and sylvite (KCl).

The U.S. Public Health Service (USPHS) standard for total dissolved solids (TDS) in drinking water is 500 ppm, although the 1974 Safe Drinking Water Act considers waters containing up to 10,000 ppm TDS as potential sources of drinking water. The four types of naturally occurring groundwater that often exceed 10,000 ppm TDS are connate water, intruded seawater, magmatic and geothermal water, and water affected by salt leaching and the products of evapotranspiration. In many areas of the United States, groundwater can be used for drinking with no pretreatment, in which case it is said to be a raw resource. In other areas pretreatment to correct hardness or color may be necessary. Where contamination by pathogenic organisms is a potential threat, chlorination may be indicated. Water from some aquifers is unusable due to its salinity or to the presence of naturally occurring toxic substances such as arsenic or radionuclides. Most of the freshwater aquifers are underlain by brackish or saline aquifers, and in general, salinity may be said to increase with depth.

Use of Groundwater

Groundwater is a major natural resource in the United States and is often more easily available than surface water. It is estimated that more than 50% of the population uses groundwater as its primary source of drinking water. It shows early estimates of population served by groundwater and surface water for drinking purposes. These estimates are from unpublished data for 1970 from the USGS. They indicate that groundwater delivered by community systems supplied 29% of the population, and an additional 19% had its own domestic wells. Approximately 36% of the municipal public drinking-water supplies came from groundwater. Of the rural population, 95% was dependent upon groundwater for drinking purposes in 1970. Of the total groundwater usage, 60% was used for public supplies and 40% for rural supplies. EPA's National Statistical Assessment of Rural Water Conditions estimated that 72.3% of all major rural household supplies was extracted from groundwater. This dependence varied regionally, with 87.5% of north-central rural households and 56.7% of western households dependent upon groundwater for water supplies. It illustrates groundwater withdrawals for rural domestic supply as a percentage of total freshwater use. This table indicates that 97% of water withdrawn for the nation's domestic rural use is supplied by groundwater. It is clear that the states west of the Mississippi River, in the area where irrigated agriculture is prevalent, depend heavily on groundwater. Arkansas, Nebraska, Colorado, and Kansas use more than 90% of their groundwater for agricultural activities. Irrigation was the principal use of all groundwater withdrawn across the country in 1980. The more humid eastern portion of the country is far less dependent on groundwater. The western states, again, are the main users of groundwater.

The most recent figures for groundwater use are from 1980 USGS data. It shows the historical trends in groundwater use as a percentage of total withdrawals. The total use of fresh groundwater showed an increase of 22% from 1970 to 1975 and an increase of only 7% from 1975 to 1980. This change in the rate of increased use may be attributable to the fact that high demand for groundwater may decrease the ease and efficiency of its withdrawal, and in some cases its quality.

Table 7.4. Groundwater Withdrawals as a Percentage of Total Freshwater Withdrawals for Rural Domestic Supply

State	Percentage	State	Percentage
Alabama	100	Montana	94
Alaska	99	Nebraska	100
Arizona	100	Nevada	94
Arkansas	100	New Hampshire	98
California	93	New Jersey	100
Colorado	36	New Mexico	97
Connecticut	100	New York	89
Delaware	100	North Carolina	100
District of Columbia	0	North Dakota	100
Florida	100	Ohio	90
Georgia	100	Oklahoma	83
Hawaii	90	Oregon	87
Idaho	96	Pennsylvania	100
Illinois	97	Rhode Island	100
Indiana	90	South Carolina	100
Iowa	100	South Dakota	94
Kansas	86	Tennessee	100
Kentucky	91	Texas	84
Louisiana	100	Utah	90
Maine	98	Vermont	85
Maryland	100	Virginia	100
Massachusetts	100	Washington	78
Michigan	100	West Virginia	95
Minnesota	100	Wisconsin	100
Mississippi	100	Wyoming	92
Missouri	74	Total national %age	97

Escalated costs resulting from these factors and the cost of the water itself have perhaps influenced users, especially irrigators, to be more efficient with groundwater use. Actual water-use figures for 1985 and 1990 (including saline groundwater withdrawals and water used for the generation of thermoelectric power) are given in Table 7.6. The historical trends in the use of groundwater and surface water, and total water use.

Table 7.5. Historical Trends in U.S. Groundwater Use as a Percentage of Withdrawals

	1960	1970	1980	1985	1990
Total Fresh Groundwater Withdrawals (bgd)	34	50	68	82	88
Public Supplies (%)	12	13	14	13	14
Rural Supplies (%)	8	6	5	5	5
Irrigation (%)	62	68	66	69	68
Industry (%)	18	13	15	14	13

Note: May not total 100% due to rounding.

In an American home, depending on the nature of the residence and the climate, exterior residential use may range from 5% to more than 150% of the interior use, averaging about 75%. This is primarily for watering lawns. Personal use of water, excluding laundry, toilet flushing, and so forth, accounts for 40% of the internal residential use and averages out to 6.3 gallons per capita per day.

The safe yield of a groundwater basin is defined as the amount of water that can be withdrawn annually without producing an undesired effect. Undesired results include depletion of the groundwater reserves, intrusion of water of an undesirable quality, contravention of water rights, the deterioration of the economic advantages of pumping, excessive depletion of stream flow, and land subsidence. Any withdrawal in excess of a safe yield is termed an overdraft. It has been suggested that the optimal yield would be a more useful concept and should be determined by an optimal groundwater-management scheme that includes and best meets economic and social objectives associated with the use of water. Under certain conditions, optimal yields may involve the mining of groundwater to depletion or complete conservation.

Increased demands on groundwater have strained the supply in certain regions, resulting in reduced artesian pressure, land subsidence, reduced spring and stream flow, and the intrusion of salt water. This kind of groundwater overdraft is occurring in the High Plains from Texas to Nebraska, parts of California, Arizona, Louisiana, Florida, New Mexico, Arkansas, Wisconsin, Illinois, and North Carolina. It has been reported that groundwater levels of the Carrizo Aquifer in south-central Texas have declined.

Table 7.6. U.S. Groundwater Use

	1985	1990
Population served by groundwater for public supplies	64,700,000	73,700,000
Total groundwater withdrawn (mgd) for public supplies	11,000	12,000
Rural use of groundwater (mgd)		
Domestic	2,700	3,300
Livestock	1,200	1,200
Total withdrawn for irrigation (mgd) to irrigate a total of 54,000,000 acres (1975) and 63,125,000 acres (1980)	57,000	60,000
Groundwater withdrawals for industrial use (mgd)		
Fresh	11,000	10,000
Saline	980	930
Water used for generation of thermoelectric power	1,390	1,600
Total withdrawals of groundwater (mgd)	85,270	89,030

Table 7.7. Allocation of Water Use In U.S. Communities

Use	Percentage
Residential	40
Commercial	15
Industrial	25
Public	5
Unaccounted for	15
Total	100

Table 7.8. Allocation of Interior Residential Water Use

Use	Percentage
Drinking and cooking	5
Bathing	30
Toilet flushing	40
Laundry	15
Dishwashing	5
Miscellaneous	5
Total	100

In the Second National Water Assessment, the U.S. Water Resources Council lists actions that could be taken before pumping becomes uneconomical in the areas with declining water levels These include locating alternate water sources, developing artificial recharge of the aquifers, relocation of water-requiring activities, and reduction of water use through better water management. Groundwater overdraft is considered to be serious when it continues indefinitely, causing the exhaustion of the resource, associated economic and social dislocations, and perhaps general weakening of a region's economy. Such potential problems are developing in California, the High Plains regions, Florida, New Mexico, Colorado, and Arizona, but many states are successfully dealing with the problem.

Profile of an Aquifer

In order to provide a picture of the importance of groundwater, it is useful to describe a specific aquifer, its characteristics and uses, and the problems that afTafect it as a natural resource. A study by the High Plains Study Council provides such data for the Ogallala Aquifer. The Ogallala is an exceptional aquifer in both extent and volume, running as it does through parts of eight states. It is an integral part of the burgeoning economy of those states, as well as of the environment. Because of its unusual size and certain other characteristics, however, it should not be viewed as a representative aquifer.

The Ogallala Aquifer is a major source of groundwater in the High Plains region, underlying a land area of approximately 33 134,000 square miles, three times the size of New York State, or about the size of California. The formation extends from southern South Dakota to northwestern Texas and transects portions of six other states: Nebraska, Wyoming, Colorado, Kansas, Oklahoma, and New Mexico. It contains an estimated quadrillion gallons, or 3.25 billion acre-feet, of water, the equivalent of Lake Huron. The aquifer fuels a $30 billion-a-year agricultural economy; irrigates over 12 million acres of farmland or 20% of the irrigated acreage in the United States; and helps support a population of nearly half a million people. Furthermore, 40% of the beef consumed in this country is fattened in the aquifer area. Such heavy exploitation of the vast groundwater resources in the Ogallala has resulted in serious depletion in some areas. These problems are now the subject of intensive study.

The Ogallala was formed during the early Pleistocene time, some two million years ago, from glacial outwash of the ancestral Rocky Mountains that consisted of gravel, sand, and finer debris that was caught up by streams of meltwater running away from the glacier. The outwash settled unevenly throughout the High Plains, giving the aquifer water-storage capacities that varied according to the depth and content of the outwash. In Nebraska, where two-thirds of the Ogallala's waters lie and where the aquifer is 1,000-1,500 feet thick, the aquifer has a considerably greater storage capacity per unit of area than it does in the South Plains of Texas, where it is less than 100 feet thick. At present, there are an estimated 2 billion acre-feet of water in storage in Nebraska compared with 350 million acre-feet in Texas and 100 million acre-feet in Oklahoma.

Annual withdrawals from the entire aquifer average 5 or 6 million acre-feet. The Ogallala's extensive groundwater resources were virtually unknown until the early part of this century. The first wells were drilled into the formation over 90 years ago, but the land was used primarily for cattle grazing and dryland wheat production well into the 1900s. The start of irrigated farming, with the use of high-capacity pumps, 34 after World War II brought rapid economic growth but also caused the water table of the aquifer to decline steadily, in some areas by as much as 100 feet, and even to the point of drying up. Gaines County, Texas, has dried up its groundwater resources entirely. The aquifer is very slow to recharge. Because of its high rate of evaporation and its high percentage of impermeable soils, the water is not replenished at the rate at which it is used. The depth of the water table nearly everywhere in the Ogallala formation is 50 feet or more below the roots of plants. Plants get the first opportunity at capturing any infiltrating rainfall, making that portion unavailable for pumping. Overall, the amount of water being overdrawn annually from the aquifer exceeds 3 million acrefeet. Annual recharge from precipitation averaging 12 to 22 inches per year, irrigation return-flows, and some stream-bed percolation is estimated to be 0.27 million acre-feet. The state of Texas has estimated that, at the current rate of use and without an imported water supply, approximately 40% of the now-irrigated acreage in the High Plains of West Texas will have to revert to dryland farming or be abandoned by the year 2000, and 60% will have to be abandoned by the year 2020.The depletion of the Ogallala began to attract public attention in the early 1970s. Members of Congress representing the Great Plains states of Colorado, Kansas, Nebraska, New Mexico, Oklahoma, and Texas encouraged the development of legislation that would mandate intensive study of the problem. A $6-million study was authorized by public law in 1976, and the High Plains Study Council was formed under contract with Camp, Dresser & McKee. The Council approved a study design with the following objectives in 1977:

1. to determine the potential development alternatives for the High Plains,
2. to identify and describe policies and actions required to carry out promising development strategies, and
3. to evaluate the local, state, and national implications of these alternative development strategies.

It is a comprehensive resource and economic-development study of the area served by the Ogallala. Major emphasis is being given to improving the water supply by local conservation and by improved practices of irrigation and agricultural management and to interbasin transfers of surplus water from adjacent basins, the Missouri River, and streams in Arkansas. Effective, comprehensive, interstate management of the 35 Ogallala groundwater resource is a further major element of the study.

Currently there is little or no public support for comprehensive, interstate management of the aquifer, although interdependencies are widely recognized. Laws in the states vary widely. Oklahoma and Texas are the only states of the High Plains with private ownership of groundwater. The groundwater in the six other states is dedicated to the people of the state, and allocation is administered by state and local officials. In all eight states, well registration or permitting systems exist.

Contamination of the Ogallala aquifer does not, to date, appear to be a major problem or concern. There has been some contamination around major population centers like Lubbock, Texas, but the depletion problem is of more immediate concern. A study by Camp, Dresser & McKee predicts that 5.1 million acres of irrigated land in six Great Plains states will dry up by the year 2020. A continuation of present trends will mean the loss of 1.6 million irrigated acres in Kansas, 1.2 million in Texas, 260,000 in Colorado, 224,000 in New Mexico, and 330,000 in Oklahoma. Twenty recommendations have been made for remedial action, but only one is an actual cure. The Army Corps of Engineers has proposed a system of huge canals that would import water from South Dakota, Missouri, and Arkansas. The cost of the project (from $3.6 billion to $22.6 billion) is prohibitive, however, and stopgap efforts like conversion to dryland farming will probably be the most immediate solutions to the depletion problem.

Uses of the Ogallala Groundwater Resources in a Few Selected States

In New Mexico the aquifer is heavily developed from southern Quay County to southern Lea County. The thickest and most productive part of the aquifer is in northern Lea County. Individual well yields in this area can be as much as 1,600 gallons per minute. In southern Lea County, where the aquifer is thinner, wells yield from 300 to 1,000 gpm, but water levels have been declining at a rate of almost 3 feet per year since 1950.

In Oklahoma the Ogallala is the most important aquifer, underlying most of the panhandle. Deposits of sand, gravel, and minor amounts of clay store high-quality 36 groundwater. The aquifer is as much as several hundred feet thick, contains more than 100 million acre-feet of available water, and supplies most of the water requirements of the panhandle. It is used to irrigate 135,000 acres, to supply water for the industrial needs of the Keyes helium plant and the natural gas industry in the panhandle, and to supply all the public and domestic needs in the area.

In Texas the aquifer formation consists of interfingered and intergraded lenses and layers of sand, gravel, silt, clay, and caliche, a crust of calcium carbonate. The High Plains is divided by the Canadian River into the northern High Plains and the southern High Plains. The northern area comprises approximately 9,300 square miles. The zone of saturation in most places is 100 to 500 feet thick. Irrigation has developed more slowly here than in the southern High Plains, which is the area of greatest groundwater development in Texas. This area includes about 25,000 square miles and has the most serious depletion problems of the entire aquifer. According to a 1985 USGS report, groundwater withdrawals in the southern High Plains have been averaging 6,500 million gallons a day (mgd). The rate of recharge has been 125 to 134 mgd. The Texas Department of Water Resources has made estimates and projections of groundwater availability in the Ogallala through the year 2030. It has determined the annual effective recharge to be 143 billion gallons. The amount of water in recoverable storage as of 1980 was determined to be 125,609.27 billion gallons. (Recoverable storage is defined as that portion of the underground reservoir capacity that, it is estimated, is physically capable of yielding water economically.) The projected annual average of groundwater availability was then determined through 2030. (Annual average of groundwater availability is defined as the estimated sustainable annual yield, or effective recharge, plus that amount of water that can be recovered from storage over a specified period of time without causing irreversible harm, such as land-surface subsidence or water-quality deterioration.)The projections are as follows:

1990	2,132.17 billion gallons
2000	2,678.33 billion gallons
2010	2,495.95 billion gallons
2020	1,959.99 billion gallons
2030	1,491.09 billion gallons

The remaining recoverable storage in 2031 is projected to be 49,696.48 billion gallons.

❑❑❑

8

Groundwater Contamination

Introduction

The mobility of our society, as well as the distribution of industry and agriculture, depends upon an available supply of clean water. Nevertheless, instances of groundwater contamination have been found in most sections of the country. For the purpose of this book groundwater contamination will be defined as the addition to water of elements, compounds, or pathogens that alter its composition.

The pollution of groundwater occurs when the concentrations of the contaminants render the water unfit for present and future uses. It shows a geographic distribution of groundwater pollution, from the Second National Water Assessment.

One of the major difficulties with groundwater contamination is that it occurs underground, out of sight. The sources of pollution are not easily observed, nor is the effect of pollution often seen until irreversible damage has occurred. There are no obvious warning signals such as fish kills, discoloration, or stench that are often early indicators of surface-water pollution. Where contamination affects pumping wells, such indicators may occur, although many commonly found contaminants are both colorless and odorless, which makes detection difficult. They occur in concentrations that, if ingested, may cause long-term chronic illness, rather than acute poisoning. Many chemicals now found in groundwater have not been tested for their effects on human health. The tangible effects of groundwater contamination usually come to light long after the incident causing the contamination has occurred. The long time lag between occurrence and detection is a major problem.

Groundwater can be contaminated by a variety of compounds, both natural and man-made. Contamination due to man has occurred for centuries, but industrialization, urbanization, and increased population have greatly aggravated the problem in some areas. Miller, Braids and Saar summarized the mechanisms of contamination and attenuation. Freeze and Cherry give a detailed account of these mechanisms, as well as of the movement of a contaminant in an aquifer.

Movement of Contaminants

A contaminant usually enters the groundwater system from the surface of the land, percolating down through the aerated soil and non-saturated zone. The root zone may extend two or three feet into the soil. Many reductive and oxidative biological processes take place in this zone that may degrade or biologically change the contaminants. Plant uptake can remove certain heavy metals; microbial fixation and other biological processes can also remove a fraction of the contaminants, the size of the fraction being dependent on the nature of the contaminant. For example, depending upon conditions, iron and manganese solubility can be affected by microbial oxidation and reduction in the soils. In the deeper geologic material that consists mainly of humus and weathered rocks, there is a lessening of such biological processes. Attenuation of contaminants in this zone may occur by surface adsorption as ions in the contaminants are attracted to the charges on the clay particles. Other contaminants may be removed by insoluble organic matter complexing, giving rise to complexed humic acids. Microbial action may influence redox potentials and cause the release of inorganic ions during decomposition. The susceptibility of different contaminants to differential attenuation varies. Present knowledge indicates that more different attenuation processes are active in the non-saturated zone than in the aquifer; and that the amounts of degradation are larger. However, many significant bio-conversions are being discovered in anaerobic subsurface environments. In fact many compounds that persist under aerobic conditions are metabolized in anaerobic conditions.

In aquifers composed of unconsolidated granular media, contaminant movement is determined by groundwater flow rates and flow paths which are governed by advection, diffusion, and dispersion, by the density of the contaminant and biological, chemical and physical interactions with the material through which it flows. Cherry states that "Contaminant migration in fractured media is determined by advection and dispersion in the fractures and by diffusive mass transfer between the flowing water in the fractures and the stagnant water in the porous but low permeable matrix."

Specific contaminants move in different ways. Liquid petroleum hydrocarbons, for example, would be expected to move downward through the aquifer; however, they may spread out on the top of the aquifer if they are present in sufficient quantities. They usually do not move very far from the point of entry because of the limiting effect of surface tension between the liquid and the water-coated grain of the unconsolidated material. However, dissolved chemicals from these petroleum hydrocarbons move with the groundwater.

Industrial liquids that are halogenated hydrocarbons have a viscosity less and a density much greater than those of water. The fractions that dissolve in water move with the groundwater. However, they may move downward, settle out in the aquitard, and (depending on the slope) may move downslope and not in the direction of the bulk flow of the groundwater.

The migration of contaminants through unconsolidated granular media is better understood than is their migration through fractured material. Unfortunately, contaminant migration through subsurface materials is a highly complex and poorly understood science. Much of the quantitative data obtained for the prediction of contaminant migration from theoretical models and laboratory data is of questionable use when applied to the field. Problems arise mainly from variable geologic conditions (heterogeneous, anisotropic, and fractured subsurface materials), and from the many types and characteristics of the contaminants themselves. Prediction of plume movement and behavior is made difficult because geologic formations are not often uniform. Layered beds and lenses may cause fingering or separation of the plume. Different geologic materials may retard or enhance the movement of the plume. The nature of the plume is further affected by the reactivity of the constituents of the contaminant. Concentrations can be altered by a variety of chemical and biochemical reactions, including adsorption-desorption reactions, acid-base reactions, solution-precipitation reactions, oxidationreduction reactions, ion pairing, and microbial cell synthesis. The plume continues to move with groundwater flow unless it is blocked and eventually reaches points of groundwater discharge such as streams, wetlands, lakes, and tidal waters. Because very little dilution takes place, concentrations of contaminants are often much higher in groundwater than in surface water.

Sources of Contamination

Sources of groundwater contamination may be divided into three main categories:

1. Natural pollution
2. Waste-disposal practices
3. Non-disposal sources due to human activities.

The degree of threat posed by these sources of contamination depends upon the concentration of the contaminant, its toxicity at that concentration, the volume of groundwater affected, the hydrogeological conditions, the uses made of the water from that particular aquifer, the population affected by such uses, and the availability of an alternate water supply.

Groundwater-pollution problems and their sources have been the object of numerous summaries. The most comprehensive listing of sources thus far is that put together by the Office of Technology Assessment. OTA identified 33 sources of known groundwater contaminants and categorized them according to the nature of their release.

Table 8.1. Sources of Groundwater Contamination

Category I — Sources designed to discharge substances	Category III — Sources designed to retain substances during transport or transmission
Subsurface percolation (e.g., septic tanks and cesspools)	Pipelines
Injection wells	Hazardous waste

Table Contd.

Hazardous waste
Non-hazardous waste (e.g., brine disposal and drainage)
Non-waste (e.g., enhanced recovery, artificial recharge, solution mining, and in-situ mining)
Land application
Wastewater (e.g., spray irrigation)
Wastewater byproducts (e.g., sludge)
Hazardous waste
Non-hazardous waste

Category II — Sources designed to store, treat, and/or dispose of substances; discharge through unplanned release
Landfills
Industrial hazardous waste
Industrial non-hazardous waste
Municipal sanitary
Open dumps, including illegal dumping (waste)
Residential (or local) disposal
Surface impoundments
Hazardous waste
Non-hazardous waste
Waste tailings
Waste piles
Hazardous waste
Non-hazardous waste
Materials stockpiles (non-waste)
Graveyards
Animal burial

Non-hazardous waste
Non-waste
Materials transport and transfer operations
Hazardous waste
Non-hazardous waste
Non-waste

Catergory IV — Sources discharging substances as consequence of other planned activities
Irrigation practices (e.g., return flow)
Pesticide applications
Fertilizer applications
Animal feeding operations
De-icing salts applications
Urban runoff
Percolation of atmospheric polllutants
Mining and mine drainage
Surface mine-related
Underground mine-related

Category V — Sources providing conduit or inducing discharge through altered flow patterns
Production wells
Oil (and gas) wells
Geothermal and heat recovery wells
Water supply wells
Other wells (non-waste)
Monitoring wells

Aboveground storage tanks	Exploration wells
Hazardous waste	Construction excavation
Non-hazardous waste	
Non-waste	
Underground storage tanks	Category VI — Naturally occuring
Hazardous waste	sources whose discharge is
Non-hazardous waste	created and/or exacerbated by
Non-waste	human activity
Containers	Groundwater-surface water
Hazardous waste	interactions
Non-hazardous waste	Natural leaching
Non-waste	Salt-water intrusion/brackish water
Open burning and detonation sites	upconing (or intrusion of other
Radioactive disposal sites	poor-quality natural water)

Changes in the Composition of Groundwater due to Natural Processes

All groundwater contains some dissolved salts. Mineralization of groundwater due to leaching is a significant source of saline groundwater in the arid areas of the country and is greatest in the areas of lowest precipitation. In areas where the water table is near the surface, evaporation and transpiration further concentrate the salts in the remaining water. Natural leaching has been a significant source of contamination in the arid southwest and south-central area comprising Arizona, California, Nevada, Utah, Arkansas, Louisiana, New Mexico, Oklahoma, and Texas. In this area, natural leaching has exceeded in importance anthropogenic sources of groundwater contamination. Many of the inorganic substances found in groundwater are largely the result of natural contamination, although human sources contribute significant amounts as well. Table 8.2 lists 37 inorganic substances that have been found in groundwater. Chlorides are a widespread contaminant. Sulfates, nitrates, fluorides, and iron are also common natural contaminants that occur in localized natural deposits, often causing the groundwater in those areas to exceed EPA standards. Radioactivity from uranium deposits has caused problems in Texas, Illinois, Iowa, Oklahoma, Nevada, Arkansas, and New Mexico. Arsenic may be a local problem in the thermal springs of the northwest region. Other examples of such localized problems occur throughout the country. Water from fault zones or of volcanic origin can contain high levels of salts or toxic chemicals. As mentioned in the section characterizing groundwater, the aquitards that confine the aquifers are not totally impermeable. A change in the composition, thickness, or continuity of the aquitard may permit leakage from one aquifer to another. The leaking of a saline aquifer into a potable aquifer may cause deterioration of the quality of water in the potable aquifer. Although natural alteration of the chemical composition of groundwater cannot be prevented, it may be fit for some uses with or without pretreatment.

Contamination due to Waste-Disposal Practices

The best estimate of the number of sources of groundwater contamination from waste disposal at the time that this report was compiled. The major sources described in this report and elsewhere will be summarized below. It should be noted that many of the numbers given are estimates and that definitive data on waste-disposal practices often were not, and are not, readily available.

Table 8.2. Inorganic Substances Found in Groundwater

Substance	Concentration (mg/l)
Aluminum	0.1-1,200
Ammonia	1.0-900
Antimony	—
Arsenic	0.01-2,100
Barium	2.8-3.8
Beryllium	less than 0.01
Boron	—
Cadmium	0.01-180
Calcium	0.5-225
Chlorides	1.0-49,500
Chromium	0.06-2,740
Cobalt	0.01-0.18
Copper	0.01-2.8
Cyanides	1.05-14

Substance	Concentration (mg/l)
Fluorides	0.12-250
Iron	0.04-6,200
Lead	0.01-5.6
Lithium	—
Magnesium	0.2-70
Manganese	0.1-110
Mercury	0.003-0.01
Molybdenum	0.4-40
Nickel	0,05-0.5
Nitrates	1.42-433
Nitrites	—
Palladium	—
Potassium	0.5-2.4
Phosphates	0.4-33
Selenium	0.6-20
Silver	9.0-330
Sodium	3.1-211
Sulfates	0.2-32,318
Sulfites	—
Thallium	—
Titanium	—
Vanadium	243.0
Zinc	0.1-240

Waste from natural and manufactured products is often stored on or beneath the land surface. In fact, burying wastes seems an ingrained cultural phenomenon. Sources of contamination are derived from all aspects of our lives, including industry, agriculture, and government.

One estimate is that between 50,000 and 75,000 chemicals are now being used and distributed through the environment and that an additional 700 to 800 are being added each year. In 1984, the OTA listed more than 200 substances that have been detected in groundwater; these include 175 organic chemicals and 50 inorganic chemicals, organisms, and radionuclides. For disposal, however, most are reacted with other chemicals, immobilized, or otherwise treated and burned.

Individual Sewage Disposal Systems

The 1980 census estimated that there were approximately 22 million septic systems serving nearly one-third of the nation's population. Annual flow to an individual septic tank ranges from approximately 49,000 to 75,000 gallons per household. The total annual flow to all domestic systems can be estimated to be from 1.08 trillion to 1.65 trillion gallons, making these systems the highest-ranking source of wastewater discharged directly into groundwater. The three methods of on-site domestic waste disposal are septic tanks, with their subsurface disposal system; the less satisfactory cesspool, commonly found in older installations where there is a deep layer of permeable soil; and finally the pit privy. Properly constructed septic-tank systems and privies permit effective treatment of human waste. Cesspools work only in coarse or highly fissured materials and have a high potential for contamination. The septic tank has several important functions. It separates solid and liquid wastes, stores solids and floatable materials, and treats aerobically both stored solids and nonsettleable material. Soils with slow percolation rates, shallow soils, soils over permeable bedrock, and soils with permanent or periodic high water tables are not suitable for the use of conventional septic tank systems.

The leach beds for septic tanks typically consist of 500 to 600 square feet of drain bed, located 18 to 24 inches below the surface of the ground. The effluent is thus distributed over a wide area, and the shallow depth of the bed permits some evaporation and some uptake of contaminants by plants. Because it is near the surface, the drain field operates under aerobic conditions, with the result that good quality water reaches the groundwater. It is important that a zone of unsaturated soil occur between the leach bed and the water table so that the effluent from the septic tank is not discharged directly into the groundwater and hence into the aquifer. The degree of potential risk posed by these systems depends in large measure upon their design and installation and upon the hydrogeology of the area. When sited correctly they operate efficiently, renovating wastewater and returning good quality water to the watershed. When operating incorrectly, they can allow pollutants to enter the groundwater. This would be particularly 62 troublesome in areas where drinking water wells tap the same aquifer in the vicinity of the leach bed.The advantages of septic tanks include the following:

- Minimal maintenance is required.
- The cost of individual or community septic tanks is less than the cost of central wastewater collection and treatment plants.
- It is a low-technology system.
- The energy requirements are low when compared with centralized wastewater-treatment facilities.

The disadvantages include:

- The potential for groundwater contamination exists when the tanks are incorrectly sited, with regard to density of the individual units or soil characteristics or both.
- The systems require proper maintenance. Failure to provide this results in system overflows or pollution of nearby wells.
- Septic-tank cleaners may severely contaminate groundwater.

The typical composition of domestic sewage is shown in Table 8.3. The constituents that pose the greatest threat to the quality of groundwater are nitrates, phosphates, heavy metals, inorganic ions (Na^+, Cl^-, $SO_4^=$, K^+, Ca^{++}, and Mg^{++}), and pathogenic organisms. In addition, toxic synthetic organic chemicals are becoming a more significant hazard due to the increased use over the last ten years of cleaning products containing such chemicals. It is difficult to show by mapping how regions are affected by sources of contamination. All contamination incidents are indicated by points on a map, but not all the groundwater in the area indicated is contaminated. Since the most important factors affecting the contamination of groundwater by sewage are the density of the individual units and the soil characteristics, the greatest potential for contamination is in the eastern region of the country. A septic-tank density of greater than 40 per square mile constitutes a region of potential contamination to groundwater.

Table 8.3. Characteristics of Domestic Sewage

Constituent	Concentration Typical Domestic Sewage
mg/l except for conductivity (micro-mhos/cm), and pH (pH units).	
Total Suspended Solids	200

Table Contd.

Constituent	Concentration Typical Domestic Sewage
Conductivity	700
Chemical Oxygen Demand (COD)	500
Biochemical Oxygen Demand (5-day BOD)	200
Total Organic Carbon (TOC)	200
pH	8.0
Alkalinity (as CaCO)	100
Acidity (as CaCO)	20
Total Phosphorus	10
Total Nitrogen	40
Chloride	50
Calcium	50
Magnesium	30
Iron	0.1
Manganese	0.1

Land Disposal of Solid Wastes

Solid wastes are defined as any garbage, refuse, or sludge from waste-treatment plants, water-supply treatment plants, or airpollution control facilities, and any other discarded materials, including solid, liquid, semi-solid, or contained gaseous materials, resulting from industrial, commercial, mining, and agricultural operations and from community activities. They do not include solid or dissolved material from domestic sewage, irrigation returnflows or industrial discharges that are point sources, or products and by-products of the nuclear industry. Solid wastes are considered hazardous when their quantity, concentration, or physical, chemical, or infectious characteristics cause, or significantly contribute to, an increase in mortality or an increase in serious illness, or pose present or potential hazards to human health or the environment when improperly treated, stored, transported, disposed of, or otherwise managed.

The solid portion of household wastes contains a high proportion of putrescible matter that is broken down by biodegradation, releasing carbon dioxide and methane gas. The leachate contains high concentrations of sulfate, chloride, and ammonia. Abundant quantities of cellulose from paper products may retard the movement of halogenated hydrocarbons by absorption. Solid commercial wastes have a similar composition to household wastes. They may, however, contain greater quantities of 64 oils, phenols, and hydrocarbon solvents. Phenols are the most resistant to biological breakdown and may be leached. Domestic and commercial wastes are serious groundwater contaminants because of the numerous dissolved constituents and the high biological oxygen demand (BOD) of may of the constituents. Table 8.4 is a summary of leachate characteristics based on 20 samples from municipal solid wastes. The table does not include synthetic organic chemicals.

The solid components of industrial wastes vary with the source of production. Cyanide wastes are produced in metallurgical operations; sulfite wastes come from paper and pulp manufacturing; mercury is a waste product in the electrical industry; and the petrochemical industry produces several solid residues ranging from polychlorinated biphenyls (PCBs) to pesticides and herbicide residues to phenol-rich tar wastes. The amount of solid waste disposed of by various manufacturing industries is listed in Table 8.5.

Other serious solid-waste contaminants can result from thermalpower generation from the burning of coal, forming fly ash which, due to its high surface area volume ratio, is fairly reactive. Leaching of the deposited fly ash may initially give rise to high concentrations of arsenic, chromium, selenium, and chloride. Another contaminating product resulting from the burning of coal is the sludge formed by the aqueous scrubbing of flue gases. Sludges typically contain cyanide and heavy metals and are of low pH unless neutralized by the addition of lime. It has been found that mixtures of sludge, fly ash, and lime rapidly set into a lowpermeability solid which leaches less readily.

Contamination occurs when precipitation infiltrates and percolates through solid wastes at poorly designed land-disposal sites, forming a leachate of dissolved minerals, heavy metals, and organic chemicals. Leachate formed at disposal sites which are located in wetlands, flood plains, or where there is a shallow water table and that are constructed without natural or artificial barriers is likely to contaminate groundwater.

Table 8.4. Summary of Leachate Characteristics Based on 20 Samples from Municipal Solid Wastes

Components	Median Value (ppm)[a]	Ranges of All Values (ppm)[a]	
Where applicable			
Alkalinity ($CaCO_3$)	3,050	0	-20,850
Biochemical Oxygen Demand (5-day BOD)	5,700	81	-33,360
Calcium (Ca)	438	60	- 7,200
Chemical Oxygen Demand (COD)	8,100	40	-89,520
Copper (Cu)	0.5	0	- 9.9
Chloride (Cl)	700	4.7	- 2,500
Hardness ($CaCO_3$)	2,750	0	-22,800
Iron, Total (Fe)	84	0	- 2,820
Lead (Pb)	0.75	<0.1	- 2.0
Magnesium (Mg)	230	17	-15,600
Manganese (Mn)	0.22	0	- 125
Nitrogen (NH_4)	218	0.06	- 1,106
Potassium (K)	371	28	- 3,770
Sodium (Na)	767	0	- 7,700
Sulfate (SO_4)	47	1	- 1,558
Total Dissolved Solids (TDS)	8,955	584	-44,900
Total Suspended Solids (TSS)	220	10	-26,500
Total Phosphate (PO_4)	10.1	0	- 130
Zinc (Zn)	3.5	0	- 370
pH	5.8	3.7	- 8.5

The types of land disposal sites for solid wastes include dumps, landfills, sanitary landfills, and secured landfills listed in increasing order of ability to protect the environment from the adverse effects of their use. Landfill regulations promulgated under the Resource Conservation and Recovery Act (RCRA) were issued in July 1982. The Comprehensive Environmental Response, Compensation and Liability Act (CERCLA) requires remedial actions at inactive sites that pose a threat to the environment and to human health. The U.S. EPA estimates that there are 93,000 landfills in the United States. Approximately.

Table 8.5. Solid Waste Disposal by Selected Manufacturing Industry Group: 1985 to 1990

Industry Group	1985	1990	1993 Total	Hazardous	Non-hazardous
All Industries	139.1	149.9	89.0	8.0	81.0
Food	12.6	14.4	9.6	0.2	9.4
Lumber and Wood	8.1	5.9	4.0	(Z)	4.0
Paper	9.1	12.3	13.7	0.1	13.5
Chemicals	38.7	43.4	18.8	3.6	15.2
Petroleum	2.0	4.9	3.6	1.5	2.1
Stone, Clay, Glass	11.3	13.3	6.2	0.2	5.9
Primary Metal	42.7	37.5	17.7	1.0	16.7
Fabricated Metals	1.9	1.9	2.0	0.2	1.7
Machinery (excluding electrical)	2.7	3.0	1.8	0.2	1.7
Electrical Equipment	1.5	2.1	1.7	0.3	1.4
Transportation Equipment	3.8	4.2	3.0	0.3	2.7

Note: In millions of short tons. Excludes recovered materials. Data include both wet- and dry-weight figures. Excludes apparel and other textiles and, beginning 1980, establishments with fewer than 20 employees. (Z) less than 50,000 short tons

75,000 of these are on-site industrial landfills and 18,500 receive municipal wastes. In 1983, another survey estimated that there are between 15,000 and 20,000 active landfills in the nation, including municipal landfills, open dumps, and nonhazardous waste facilities, operating, abandoned, or closed.

Hazardous wastes may be disposed of in a landfill, dump, or impoundment. EPA requires that site locations and amounts and types of wastes stored be reported, under CERCLA by owners and operators of inactive sites, and under RCRA by present owners and operators of active privately owned industrial sites. About 199 active hazardous-waste landfill facilities and approximately 20,000 abandoned hazardous-waste sites are known to exist across the country.

The amount of solid waste generated in this country has been estimated to be enough to cover 400 acres of land to a depth of 10 feet each day. It is estimated that in 1980, 3.74 pounds of solid waste were generated per person per day, amounting to 155.6 million tons of municipal wastes nationwide for the year. OTA also estimates that 30 million tons of solid wastes generated by utilities and 1 million tons of municipal sludge are disposed of in landfills per year. Between 40 million and 140 million wet tons of nonhazardous industrial solid wastes are disposed of annually. In 1981 hazardous-waste facilities received 14.7 billion gallons of both liquid and solid hazardous wastes.

Collection, Treatment, and Disposal of Municipal Wastewater

The collection, treatment, and disposal of municipal wastewater may be a problem in urban areas. Municipal wastewaters include domestic wastewater, storm water and its associated debris, and industrial wastes. They are collected by sanitary sewer systems and transported to treatment sites. Approximately 170 million Americans were served by sewer systems in 1984.

Leaks may occur in the system due to design, age, disruption by tree roots, seismic activity, or poor construction. The lagoons and ponds used in wastewater treatment operate under anaerobic, aerobic, or facultative conditions. The total volume of sewage handled by one of these plants is estimated to be approximately 0.667 mgd. Lagoons are often constructed without adequate seals, thus promoting leakage under certain geologic conditions and posing a threat to groundwater. Another possible route of groundwater contamination is by the land disposal of treated wastewater, which could alter the chemical constituents of the natural groundwater. Land disposal methods include agricultural irrigation, rapid infiltration ponds, overland runoff, and discharge into dry stream beds and ditches. Methods involving spraying onto land are not suitable for use in freezing winter months. The geographic distribution of various densities of population that are served by municipal sewagetreatment facilities that utilize the land disposal of effluent.

Industrial and Other Wastewater Impoundments

Under the Safe Drinking Water Act, EPA undertook a national survey of surface impoundments, which was carried out from 1986 to 1988, and the interim report was published in 1988. The final report, released in 1993, contains data from 1988 through 1990. These reports define surface impoundments as depressions in the land (pits, ponds, lagoons, and pools) containing liquid, semi-solid, and solid wastes. In the final report EPA stated that the number of sites located in the United States totaled approximately 80,263 and contained more than 180,000 impoundments. Thirty-one percent of the sites were oil 68 and gas brine pits, 25% municipal sites, 18% agricultural sites, 15% industrial sites, and 9% mining sites. Table 8.6 shows EPA's assessment of these sites. At the time of the survey it was estimated that approximately 70% of the industrial sites, 84% of the agricultural sites, and 78% of the municipal sites remained unlined. More than 98% of the sites were located in areas with proximity to potential water supplies. Approximately 15% of the sites (excluding oil- and gas-related facilities) contained hazardous wastes. New Mexico was estimated to have the largest number of sites, with 12,901, and Rhode Island the fewest, with 44. Table 8.7 shows the total number of impoundments and impoundment sites by state. The regulations under RCRA may have affected the validity of these estimates, but no recent figures are available.

Table 8.6. Summary Statistics for Located Active Surface Impoundment Sites

Category	Located Sites		Assessed Sites		Located Impoundments	
Industrial	11,760	(15%)	8,662	(28%)	27,912	(15%)
Municipal	19,746	(25%)	10,822	(34%)	37,185	(21%)
Agricultural	14,850	(18%)	6,646	(21%)	19,437	(11%)
Mining	7,364	(9%)	1,552	(5%)	25,038	(14%)
Oil & Gas Brine Pits	24,990*	(31%)	3,354	(11%)	65,488	(36%)
Other	1,553	(2%)	350	(1%)	5,913	(3%)
Total	80,263		31,386		180,973	

Note: SIA site numbers for the mining and oil & gas brine pit sites are usually related to lease or field data, not to actual ownership, and should not be referred to as the actual number of legal sites. The number of located impoundments would be a closer approximation for these two categories. *Located sites:* Total number of facilities identified in the inventory. *Assessed sites:* Total number of facilities evaluated in the inventory. *Located impoundments:* Total number of impoundments identified in the inventory. The number is larger than located sites since many facilities had more than one impoundment.

Table 8.7. Estimated Number of Located Impoundments and Impoundment Sites for All Categories, by State

State	Active		Abandoned	
	Sites	Impoundments	Sites	Impoundments
Alabama	1,063	1,696		
Alaska	129	179	27	31
Arizona	641	1,615	20	48
Arkansas	861	6,806	12	24
California	1,750	7,577	31	101
Colorado	1,267	3,631		
Connecticut	305	758	83	188
Delaware	52	131		
Florida	3,644	5,610		
Georgia	1,363	1,640	65	70
Hawaii	94	296		
Idaho	253	538		
Illinois	4,471	6,677	591	745
Indiana	1,897	2,678		
Iowa	1,784	2,539		
Kansas	1,970	3,328		
Kentucky	640	922		
Louisiana	984	2,804	49	77
Maine	132	319	41	133
Maryland	443	836	18	25
Massachusetts	341	2,211	38	208
Michigan	2,187	3,505	76	135
Minnesota	1,952	2,733	135	160
Mississippi	1,306	1,667	52	58
Missouri	3,657	4,683	507	507
Montana	668	1,266		
Nebraska	1,632	3,254		
Nevada	216	541	10	382
New Hampshire	152	274	2	4
New Jersey	311	896	46	129

Table Contd.

State	Active Sites	Active Impoundments	Abandoned Sites	Abandoned Impoundments
New Mexico	12,901	17,746	39	116
New York	1,070	1,837	40	52
North Carolina	709	1,051	26	30
North Dakota	667	1,168		
Ohio	3,287	16,537		
Oklahoma	2,935	4,538		
Oregon	370	714		
Pennsylvania	7,346	34,224	234	627
Rhode Island	44	133	3	13
South Carolina	1,868	2,259		
South Dakota	606	926		
Tennessee	746	1,325	59	151
Texas	3,672	10,740	170	365
Utah	211	2,345		
Vermont	214	274	10	17
Virginia	1,212	1,975	30	62
Washington	403	1,049		
West Virginia	688	1,861	66	186
Wisconsin	1,022	1,720	49	82
Wyoming	1,146	1,738		
Total	79,818	180,328	2,519	4,726

Impoundments which are used for the treatment, storage, or disposal of wastes are natural or man-made, and may or may not be lined. Discharging impoundments are designed to discharge regularly into bays, oceans, lakes, or streams. The liquid in nondischarging impoundments is lost by evaporation or by seepage into the soil. The types of wastes received by these sites range from sewage wastes, industrial wastes including those from air scrubbers, cooling-tower blow-down, and ash residues and oiland gas-extraction wastes to animal-feedlot and other agricultural wastes. Of these, the largest users of surface impoundments are the oil and gas industries. Evaporation ponds were used extensively in the South for oil-brine disposal. Evaporation ponds usually lose more fluid by infiltration than by evaporation, especially in the more humid regions of the country. EPA promulgated regulations for hazardous-waste surface impoundments in July 1982. In addition, state regulations exist in most oil-producing states.

The chemical character of the waste varies enormously and may include suspended and dissolved solids, pathogenic organisms, oil and grease, detergents, heavy metals, and toxic organic chemicals. It may have a high biological oxygen demand (BOD), chemical oxygen demand (COD), and total organic carbon (TOC). The pH also varies. Not all surface impoundments contain hazardous wastes. Some are used as holding ponds prior to the treatment of the fluids. Because the chemical character of the wastes varies, only a minimal number of impoundments would hold all of the contaminants listed. The type of groundwater contamination known to occur as a result of leaking impoundments reflects the nature of the fluids in the impoundments. Dissolved materials, both inorganic and organic, move into groundwater by direct seepage, while solids may be leached by precipitation or waste fluids. As with all sources of contamination, the sorptive capacity and low permeability of some soils may slow or impede the movement of some of the contaminants.

Land Spreading of Sludge

When the wastewater from municipal and industrial wastes is treated, a residue of sludge remains. Beginning in November 1981, all facilities disposing of sludge considered hazardous on land were required to have a groundwater-monitoring system in place under Resource Conservation and Recovery Act regulations. In 1982 there were at least 2,463 publicly owned treatment facilities applying liquid and thickened sludge on land surfaces, and an additional 485 facilities in operation or under construction using sludge spray irrigation practices. Of the 6.8 million dry tons of sludge produced by municipalities in 1982, 24%-29% was spread directly onto crops. The purpose of land treatments is to biodegrade the organics and to immobilize the inorganics. The potential for contamination exists, however, since the chemical constituents and viruses in sludge may be leached by precipitation after the sludge is spread on the land. Sludge from manufacturing industries may contain either useful agricultural chemicals or toxic compounds, and it varies from degradable to very refractory. Sludge is stabilized and may be dewatered before land application. The higher the proportion of clay and organic colloids in the soil, the greater the capacity for the heavy-metal ions present in sludge to be immobilized. The main hazardous constituents of the sludges include heavy metals and organic chemicals, such as dyes, inks, oils, pesticides, detergents, organic solvents, and polynuclear aromatic hydrocarbons (PAHs); organic chemicals are typically trace contaminants. Sludges from the municipal treatment of wastewater or the biotreatment of industrial wastewater include biological components such as bacteria, viruses, fungi, algae, protozoans, rotifers, and other parasites such as worms and flukes. The constituents that may be found in various types of industrial wastes are enumerated in the 1977 Report to Congress. As of 1985, EPA reported seven states that had experienced groundwater contamination due to land spreading of sludges.

Land farming of sludge has been used for nearly 30 years by the petroleum industry. From its research on this method of disposing of oil sludge, the Solid Waste Management Committee of the American Petroleum Institute concludes that if the area farmed is carefully matched to the amount of organic material that must be degraded, the type of land farming practiced, and the degradation rate of the sludge, then the underlying soil and groundwater will not be significantly contaminated.

Brine Disposal Associated with the Petroleum Industry

Groundwater contamination associated with oil production has been documented in at least 21 states. During 1983 the total production of crude petroleum was more than 3 billion barrels, valued at more than $83 billion. The principal oil-producing states in 1983 were Texas, Arkansas, Louisiana, and California. Brief mention of the disposal of brines by the use of surface impoundments was made above. Oil production is usually accompanied by the production of saline wastewater in amounts that vary with production procedures. One estimate is that an old well may produce 100 barrels of brine for each barrel of oil. Others estimate that the production of 8 million barrels of crude oil produces 30 million barrels of saltwater and that the ratio of brine to crude oil recovered is 10:1. Still others estimate a ratio of 4:1 brine to oil.

In the early days of oil and gas production, the brine was often disposed of in unlined pits. The extensive use of such "evaporation" pits that leaked their contents has caused groundwater contamination in the oil-producing regions. Use of unlined pits has been completely eliminated in some states and partially eliminated in others. For example, unlined pits were banned in Texas in 1969. Other states permit the use of small pits in remote areas if they pose no threat to groundwater.

Present practices for brine disposal include injection of the brine into deep underground formations that are deemed unsuitable for other purposes and reinjection of brines into oil-producing formations to enhance oil recovery. These injection practices, if not properly controlled, pose a potential threat to groundwater.

Past oil production practices, including well drilling, completion, and abandonment practices, also were not as strictly controlled as they are now. Some wells were improperly equipped, and abandoned wells were improperly plugged, thereby providing a potential for groundwater contamination. In 1983 abandoned production wells numbered around 1.2 million. The need to protect groundwater was recognized, however, and state oil and gas regulatory agencies now have regulations concerning brine injection, well completion, and plugging practices. The Interstate Oil Commission sponsored a study in 1978 that concluded that in Texas, Arkansas, Louisiana, New Mexico, and Oklahoma, reinjection of brines had resulted in fewer instances of groundwater contamination than had the use of evaporation pits. It therefore believes that the practice does not pose a major threat to the groundwater supplies.

Most oil-field brines are corrosive, but the corrosion rate of old and new well casings is unpredictable, and frequent monitoring and surveillance is therefore carried out. Although brine is the main threat to the quality of groundwater related to oil production, minor threats are also posed by the oil and gas themselves, by drilling fluids, by chemicals used in treating wells, by corrosion inhibitors, and by other chemicals. Oil drilling may entail drilling through freshwater aquifers, thus raising the potential for contamination.

Disposal of Mine Wastes

Mining creates conditions and products conducive to groundwater contamination. The recharge and movement of groundwater may be affected by the mining process. Domestic mining for minerals other than organic fuels was more than a $20-billion industry in 1983. Domestic mining for minerals other than organic fuels was a $10-billion industry in 1974. Coal has enjoyed a comeback as an energy source due to the goal of increased self-sufficiency in energy, and it is estimated that between one-third and one-half of the world's total reserves are found in the United States. The distribution of coal fields in the conterminous United States. Techniques of surface and underground mining pose different threats to groundwater quality, although the threat from waste disposal from the two is similar. Although such mining changes the environment, good reclamation may stabilize some of these changes without causing serious damage to the groundwater. Sand, clay, gravel, and stone quarrying are not considered a serious threat to groundwater quality. Threats from the remaining categories of mines include those posed by leachate from tailings piles and drainage from strip, surface, and underground mines. Dewatering of mines, involving a regional lowering of the water table, may have three results: a decrease in the amount of groundwater supplied, a lowering of the water level below the intakes of productive wells, and the oxidation of exposed minerals. Coal is often found in conjunction with iron pyrites (FeS_2) and other sulfides, and the oxidation of these sulfides and the presence of percolating waters produce sulfuric acid and an increase in Fe^{++} and $SO_4^{=}$, leading to acidic groundwater, a serious problem in the coal-mining areas of Pennsylvania and Appalachia. It is very costly either to neutralize acid mine drainage or to backfill abandoned mines in order to prevent the problem. Sealing the mines with airor watertight seals has not proved very successful. A report by the National Academy of Sciences summarizes these and other effects that coal mining may have on groundwater quality.

Metal mining produces drainage waters that have lower concentrations of iron and sulfate than do colliery waters, but that have higher concentrations of dissolved heavy metals due to the lower pH caused by the oxidation of metal sulfides. Wastewater from metal mining may also contain organic flocculants from the on-site processes of screening and dressing the ores. The wastes from such operations, which can be both toxic and radioactive, may contaminate groundwater. The tailings ponds used in the waste-disposal operations of mines may contribute contaminated leachate.

The waste products of quarrying for stone, lime, gypsum, and the like are generally inert, and their only contributions to the water that percolates through them is the possibility of an occasional increase in suspended-solids if crushed material is present. Quarrying does, however, increase the potential for local groundwater contamination in other ways. The removal of soil reduces the possibility of an attenuation of any spills or leakages that may occur in the area and increases the potential for contamination, especially if the bedrock is fractured.

Because of the serious nature of surface-water contamination by mining activities, more attention has been given to it than to groundwater contamination, although aquifers in regions with a long history of mining have been written off as sources of water supply. Since then coal mining on a large scale has been initiated in the West, particularly in North Dakota, Wyoming, and Montana. The EPA estimated that 3.6 million tons of acid may be generated annually from the approximately 200,000 acres used across the nation for disposal of coal-mining wastes. The EPA also estimated that 10% of the acid enters groundwater each year. Mining operations generate approximately 2.3 billion tons of total waste material annually, including radioactive and nonradioactive waste piles and tailings.

The most serious threat to groundwater is posed by the metal mines, particularly uranium and copper mines, which produce waste containing dissolved toxic materials (e.g., arsenic, sulfuric acid, copper, selenium, and molybdenum), as well as radioactive materials.

Deep-Well Disposal of Liquid Wastes

The injection of a variety of liquid wastes into deep wells to avoid having them contaminate the biosphere was becoming a widely adopted waste-disposal practice in the late 1980s, mainly because there were stringent regulations governing disposal into surface waters. Such wells are less widely accepted now in certain states. It has been estimated that there are more than 221,000 wells in the United States injecting liquid wastes underground: approximately 280 are used for hazardous wastes; 140,000 are used either for brine disposal or for the injection of fluids used in enhanced oil-recovery process; 40,000 are used for agricultural drainage, urban runoff and sewage disposal; and at least 12,000 are in operation for solution mining.

Because of the hazardous nature of the injected fluid often radioactive, toxic chemical, petrochemical, and pharmaceutical wastes that are difficult to treat deep-well disposal may pose a great threat if improperly practiced. Eighteen states have already experienced contamination through injection wells. The Underground Injection Control Program of the Safe Drinking Water Act, which prohibits direct injection into drinkingwater aquifers, and regulations issued under the Resource Conservation and Recovery Act provide such controls. All injection wells are in the depth range of 660 to 13,200 feet, most being between 990 to 6,600 feet. Injection zones are usually located in sandstone, basalt, and carbonate rocks. Injection occurs under pressure, and injection rates vary from 500 to 370 gallons per minute. Injection causes hydrodynamic changes in the aquifer, including the formation of a mound in the potentiometric surface extending in the direction of the regional flow in the aquifer. This is essentially the reverse of what happens when a well is pumped in a confined aquifer. Until 1976, documented cases of disposal-system failure were rare, but they may increase if underground injection becomes more common and occurs over longer periods of time. Underground injection in the vicinity of old unplugged wells, the locations of which are often not known, may lead to the upward leakage of the wastes. These sources estimate that there may be more than one million unplugged, unlocated wells in North America that were used for purposes other than disposal. Another hazard associated with the practice of waste injection but only indirectly linked to groundwater contamination is the triggering of earthquakes due to increased pore-water pressures along fault lines. The most publicized occurrence, at the Rocky Mountain Arsenal in Colorado, occurred after injection of chemical-warfare waste, using a 12,000foot well, into fractured pre-Cambrian rocks. The earthquakes thus caused stopped when injection ceased. Other problems may arise if an injection well becomes clogged with injected suspended solids, corrosion products, or precipitation forming between injection fluid and connate water in the aquifer. Insufficient treatment of injected water may lead to bacterial clogging, and swelling of the mineral constituents of the geologic formation.

The Report to Congress stated that injection wells can cause groundwater contamination through the following mechanisms:

- direct injection into drinking-water aquifers
- leakage into potable aquifers due to well-construction faults or failure
- leakage through confining beds by hydraulic fracturing or insufficient thickness
- displacement of saline water into potable aquifers
- migration to potable water zone of the same aquifer.

The report concluded, however, that properly constructed, sited, and monitored waste injection wells can be operated with little danger of groundwater contamination.

Disposal of Wastes from Animal Feedlots

Groundwater contamination by the manure from high-density animal-feeding operations is a relatively new phenomenon. In addition to the threat to groundwater from the vast quantities of manure thus generated, there is also a threat from the food additives, such as hormones and antibiotics, which the manure may contain. It was claimed that feedlots pose no danger of groundwater contamination in the southeastern states due to the soil's ability to denitrify the wastes. Feedlots for poultry and hog farms in Arizona, California, Nevada, and Utah were thought to lead to water-quality problems from bacteria, viruses, nitrate concentrations and with general color, taste, and odor. The main beef-raising areas, however, are the Corn Belt and the High Plains. Until the 1950s and 1960s, beef was raised on pasture, and the wastes were easily assimilated naturally and posed no problem. The need for more meat at reasonable cost led to the establishment of feedlots, with capacities of between 1,000 and 50,000 head of cattle. During the four to five months each animal spends in the lot, it produces 0.5 tons of manure (dry weight). This can lead to high nitrate concentrations in the groundwater under some conditions. Where the feedlots are located in areas with a deep water table, the risk of contamination is minimized. The main poultry-rearing region is the Mid-Atlantic and South, in Delaware, Maryland, Virginia, Arkansas, Georgia, and Mississippi, for hogs it is the Midwest, and for sheep the Far West. Of the potential contaminants in manure, nitrate is the most important, as it is soluble in water and its concentration remains unchanged while passing through the soil. The other contaminants from manure bacteria and phosphate are highly attenuated by the soil and pose less threat. Feedlot management, including stocking rate, density, and manure removal, plays an important role in the protection of groundwater quality. Heavy manure accumulations may produce an impermeable mat, which in turn produces anaerobic conditions that favor denitrification. Thus nitrate is volatilizied, and little infiltration takes place, minimizing the nitrate-contamination potential. Concentrated animal feeding operations are regulated by the Federal Water Pollution Control Act Amendments of 1992 and may require a permit as a point source under the National Pollutant Discharge Elimination System (NPDES).

Contamination from Radioactive Sources

Radionuclides from natural sources, radioactive fallout, nuclear testing and accidents, nuclear fuel cycling, commercial and industrial products and wastes pose a threat to groundwater quality. Almost all groundwater contains some amount of naturally occuring radioactive substances. Types and levels vary geographically and depend on local geology. The highest concentrations of natural radiation of groundwater are often found near granite and phosphorus formations.

Radioactive wastes from the production and generation of nuclear fuel and radioactive materials can be categorized into five types: spent fuel, high level wastes, transuranic wastes, low level wastes and uranium mill tailings. Quantities of these waste types generated as of 1983 and projected quantities are presented in Table 8.8.

Spent nuclear fuel is irradiated fuel resulting from nuclear power plant operations and is highly radioactive. Most of these wastes are presently being stored in pools at reactor sites until underground repositories are found for permanent disposal.

High level wastes result from the initial processing of spent fuel and must be stored carefully in specially constructed facilities. Presently these wastes are being temporarily stored at four utility and defense sites awaiting permanent disposal. Possible sites for a high level repository are being considered in Groundwater contamination by high level radionuclides has been reported in a few instances, largely due to the improper storage of wastes. For example, in the early 1950s, one hundred gallons of high level radioactive wastes leaked from a tank at the Savannah River defense storage site, contaminating some groundwater. In 1973 a 115,000-gallon leak of high level radioactive wastes occurred at the Hanford Nuclear Reservation. No groundwater contamination resulted, although the soil beneath the site remains radioactive. Another extensive leak of high level radioactive wastes occurred in 1956, spilling 450,000 gallons of high level radioactive wastes in the ground at Hanford, Washington, but with no serious contamination resulting.

Table 8.8. Current and Projected Quantifies of Radioactive Waste and Spent Fuel

Type of Waste	Volume of Waste (1000 m^3)		
	1983	2000	2010
High Level Waste			
Commercial	2.3	.436	3.17
Defense	304	294	269
Transuranic Waste			
Commercial	—	4.6	49.4
Defense	246	340	397
Spent Fuel			
Commercial	4.626	19.378	33.258
Defense	0	0	0
Lower Level Waste			
Commercial	1,020	3,393	5,415
Defense	2,060	3,720	4,710
Inactive Uranium Mill Tailings	14,314	23,700	23,700
Active Mill Tailings	96,500	188,800	280,300

(DOE/Defense tailings are about 35% of these totals.)

1. No reprocessing.

2. Found primarily at inactive uranium mills located in western United States. DOE responsible for stabilization and control of mill tailings in safe and environmentally sound manner under the Uranium Mill Tailings Radiation Control Act of 1978. Anticipated cleanup completion in late 1980s.

3. Located at the 16 active licensed uranium mills operating in 1983, all in the western United States.

Transuranic wastes contain alpha-emitting radionuclides of atomic numbers greater than 92. They result primarily from fuel reprocessing and the manufacture of plutonium containing products. These wastes take a long time to decay and require long-term isolation for proper disposal. Presently, there are a total of 7 sites where transuranic wastes may be deposited.

Low level wastes exist in all phases and consist of a wide range of waste material. Most of the radionuclides in low level wastes are short lived and have low radioactivity, but are produced in potentially hazardous amounts. These wastes are generated by hospitals, laboratories, industrial plants, nuclear power plants, and government and defense laboratories and reactors. In the past, low level wastes were either buried in trenches at government sites or dumped out at sea. Now all commercially produced low level wastes are permanently disposed of in one of the three commercial shallow nuclear burial facilities located in Beatty, Nevada, Barnwell, South Carolina, and Hanford, Washington, the major site being in Hanford. The Department of Energy and the Department of Defense also maintain a number of sites operated by private contractors for low level disposal. Shallow burial of low level wastes present a great potential for contamination. Groundwater may erode the structure of the burial site thus providing avenues for contamination to seep. A recent study found that water has come into contact with wastes in burial trenches at 6 of the 11 major low level governmental and commercial burial facilities.

Uranium mill tailings are earthen residues, often in the form of sand, left after uranium is extracted from ores. Disposal commonly takes place in shallow burial grounds located near refineries. Remedial action at inactive mill tailing sites are to be conducted by the Department of Energy under the Uranium Mill Tailings Radiation Control Act. According to the League of Women Voters twenty-four abandoned disposal sites are scheduled for clean up by DOE in Arizona, Colorado, New Mexico, North Dakota, Oregon, Pennsylvania, South Dakota, Texas, Utah, Washington, and Wyoming. Of the 28 active uranium mills licensed in the US, 21 reported some degree of groundwater contamination. Remedial action has been taken at 16 of the sites so far.

Other Sources of Contamination

Accidental Leaks and Spills

Accidental leaks and spills can occur because of breaks in pipelines, leaks in underground storage tanks, spills from tanker trucks, and other accidents involving the transport and storage of materials. They commonly occur at airports, industrial sites, along highways, railroad sites, gas stations, and refineries.

Small hydrocarbon spills may be absorbed by the unsaturated soil zone, but large ones can reach the water table aquifer and float on the surface. Very small concentrations of petroleum products will render groundwater unfit to drink because of its objectionable odor and taste and also pose a threat to health.

Other chemicals have toxic properties. Ammonia, for example, increases the nitrification of groundwater, and acids speed up the solution of heavy metals and soil solids. The U.S. General Accounting Office reported that leaks from underground storage tanks (USTs) have been reported in all 50 states; many states claim that this source is their leading cause of underground contamination. There are an estimated 2.5 million underground storage tanks across the U.S., containing more than 25 billion gallons of non-hazardous liquid products, and an additional 2,031 hazardous storage and treatment tanks containing an estimated 13.8 billion gallons of substances.

Data was collected at the state level on the numbers of release incidents for EPA's report *Summary of State Reports on Releases for Underground Storage Tanks*. State files identified 12,444 release incidents that had occurred across the country as of 1984. Even though this data represents the most comprehensive currently available information on leaking USTs, these numbers cannot be regarded as representing all storage tank releases. Differences in state enforcement and reporting procedures have resulted in inherent biases in the data. The results of the study showed that the greatest number of incidents occurred in the Northeast, Mid-Atlantic, and Great Lake states. This is probably due to the more corrosive conditions of the soils, the higher tank population, and perhaps more efficient procedures for reporting and documenting leaks. Soils became contaminated in 68% of the cases, and groundwater contamination resulted from 45% of the releases. It was found that approximately 65% of the incidents originated from retail gasoline stations, and only 3% of the releases involved chemicals other than petroleum fuels. Ninty-five percent of the leaks were reported to have taken place from operating facilities. The mean tank-leakage age was found to be 17 years for all of the tanks. Eighty-one percent of the tanks were steel and the remaining 19% fiberglass. The most commonly reported cause for release was corrosion and structural failure.

The NAS estimated that 16,000 spills occur annually during transport operations. In 1981 alone, 10,072 spills from both oil and hazardous chemicals, totaling 19.6 million gallons were reported. The Department of Transportation reported that more than 9,000 spills of hazardous materials occurred during transportation in 1981 and 6,500 in 1982. Emergency measures to deal with spills often consist of flushing the spilled compound away, thus permitting or even assisting in the contamination of groundwater. Flushing usually reduces the immediate threat of fire and explosion. Such practices are more likely to create surface-water pollution than groundwater pollution, as most storm sewers and ditches flow into surface water. If the spill is large enough and is not properly cleaned up, however, groundwater contamination can result. Under the Clean Water Act, Spill Prevention Countermeasure Control (SPCC) plans have been required of every facility since 1995 for spills to the ground. The National Contingency Plan, under CERCLA, effective December 1992, details cleanup procedures for spills of oil and hazardous materials on land, into surface water, or into groundwater. In addition, releases of quantities of oil or hazardous material in excess of stated amounts must be made known to the National Response Center. Many of the industrial trade associations have researched cleanup and containment and have published manuals for example, the American Petroleum Institute manual for the cleanup of underground spills.

Pipelines are used to transport a variety of fluids, but the most common one is petroleum. In 1972 there were 174,000 miles of petroleum pipelines that transported a total of 8.5 billion barrels of petroleum products. In 1976 the same amount of pipeline transported 9.63 billion barrels of petroleum products. There were 308 interstate pipeline accidents, involving a loss of about 245,000 barrels of liquids. Pipeline failure due to external corrosion caused 33% of these accidents, and damage to pipeline from excavating machines caused an additional 22%. In 1980 there were 275 accidents; in 1981, 198 accidents. The decrease in incidents may be due to extensive regulations from the Department of Transportation regulating transportation of liquids by pipeline. Of the 4,112 accidents that occurred between 1968 and 1981, 1,372 were due to corrosion and 1,101 to pipeline 83 ruptures. No information was given about the number of barrels involved in the incidents in 1980 and 1981.

In some regions deliberate leakage and spillage of contaminants, so-called midnight dumping, has occurred. This source of contamination, which is a criminal offense, is considered to be a major problem in New England.

Agricultural Activities

Irrigation return-flow; the use of pesticides, fertilizers, and manure; and changes in vegetative cover have all been known to cause changes in groundwater quality. Contamination of groundwater by nitrates is one of the largest sources of non-point pollution in the United States. The USGS has collected nitrate data over the past 25 years from approximately 124,000 wells nationwide. Its data revealed areas of groundwater that exceed EPA limits for nitrate in almost every state. The study suggests that a large contributing factor to elevated levels of nitrates are agricultural activities such as the use of fertilizers, irrigational practices, dryland farming, and livestock wastes. Other contributing factors mentioned are septic systems, land disposal of wastes, industrial wastes, and various natural sources. An EPA study of rural water conditions, however, found that only 2.7% of the nations rural wells had nitrate levels exceeding EPA's limits. In the North Central region, 5.8% and in the Northeast region, 0.3% were reported in violation of standards. Hallberg determined that since the 1960s, an increase in the use of nitrogen fertilizers has been paralleled by a similar increase of nitrates in groundwater. Fiftyfour million tons of commercial fertilizer were used between 1980 and 1981, 48.7 million tons between 1981 and 1982, and 42.3 million tons between 1982 and 1983. Nitrogen accounts for 6.1% to 20.4% of these weights. Much of the applied nitrate is not utilized by crops and is subsequently lost through leaching.

Table 8.9. Summary of Nitrate-Nitrogen Concentrations in Groundwater

State	No. of Wells Sampled	% of Wells with NO_3-N 10 mg/l
Alabama	244	0.0
Alaska	1,305	2.4
Arizona	4,164	13.9
Arkansas	2,436	3.9
California	2,732	10.1
Colorado	5,492	5.7
Connecticut	348	2.3
Delaware	165	9.1
Florida	3,140	2.0
Georgia	1,137	0.5
Hawaii	164	0.0
Idaho	1,806	1.7
Illinois	359	8.4
Indiana	650	1.4
Iowa	4,088	5.0
Kansas	1,140	20.0
Kentucky	3,227	4.2
Louisiana	3,177	0.6
Maine	147	2.0
Maryland	1,215	6.8
Massachusetts	414	1.2
Michigan	1,108	1.1
Minnesota	1,655	9.3
Mississippi	1,701	0.2
Missouri	2,165	2.1
Montana	2,821	3.8
Nebraska	2,326	9.3
Nevada	465	0.9
New Hampshire	69	1.4
New Jersey	1,385	1.4
New Mexico	4,685	2.9
New York	2,491	11.0
North Carolina	908	0.8

State	No. of Wells Sampled	% of Wells with NO_3-N 10 mg/l
North Dakota	7,387	4.6
Ohio	339	2.6
Oklahoma	1,724	11.8
Oregon	685	1.2
Pennsylvania	4,326	5.9
Puerto Rico	79	2.5
Rhode Island	171	36.3
South Carolina	557	0.7
South Dakota	1,996	6.7
Tennessee	109	0.9
Texas	36,196	9.4
Utah	3,301	2.0
Vermont	73	1.4
Virginia	762	0.8
Washington	1,158	4.3
West Virginia	954	0.5
Wisconsin	2,727	3.6
Wyoming	1,477	3.8
Total	123,656	6.4

Less nitrate-nitrogen is available for leaching in the loamy finesand soils of subirrigation systems than in furrow and sprinkler systems. Thus, subirrigation systems provide a method for reducing solute concentrations in irrigation return-flow. Irrigation only contributes further to nitrate leaching. Less than 1% of the total cropland in the Northeast is irrigated; as a result, nitrate contamination is not that much of a problem. In the West, however, due to the highly saline soils, irrigation often exceeds soil requirements. Dryland farming in the northern Great Plains does not utilize irrigation, and crop plants grown in the area often have lower evapotranspiration rates; therefore water may move down through the soil picking up salts and thus contaminating the groundwater. In some cases this salty water produces saline seeps. In 1983, 12 states experienced groundwater contamination from pesticides applied to agricultural land. In 1984, a total of 12 pesticides were discovered in the groundwaters of 18 states. By 1985, 23 states discovered groundwater contamination from 17 different pesticides.

Although agricultural operations alone account for 69%-72% of all pesticide use, contamination by pesticides can also occur through spills, leaks, lawn and garden application, and disposal practices.

Table 8.10. Pesticides Found in Groundwater as a Result of Normal Land Applicationa

Pesticide	Use	State(s)	Typical Positive, ppb
Alachlor	H	MD, IA, NE, PA	0.1-10
Aldicarb (sulfoxide & sulfone)	I, N	AR, AZ, CA, FL, MA, ME, NC, NJ, NY, OR, RI, TX, VA, WA, WI	1-50
Atrazine	H	PA, IA, NE, WI, MD	0.3-3
Bromacil	H	FL	300
Carbofuran	I, N	NY, WI, MD	1-50
Cyanazine	H	IA, PA	0.1-1.0
DBCP	N	AZ, CA, HI, MD, SC	0.02-20
DCPA (and acid products)	H	NY	50-700
1,2-Dichloropropane	N	CA, MD, NY, WA	1-50
Dinoseb	H	NY	1-5
Dyfonate	I	IA	0.1
EDS	N	CA, FL, GA, SC, WA, AZ, MA, CT	0.05-20
Metolachlor	H	IA, PA	0.1-0.4
Metribuzin	H	IA	1.0-4.3
Oxamyl	I, N	NY, RI	5-65
Simazine	H	CA, PA, MD	0.2-3.0
1,2,3-Trichloropropane	N (impurity)	CA, HI	0.1-5.0

† Total of 17 different pesticides in a total of 23 different states.

E ? H = herbicide; l = insecticide; N = nematicide

EPA believes that with increased groundwater monitoring more cases of contamination involving additional agricultural chemicals will be discovered. The pesticide dibromochloropropane (DBCP), now banned in the United States, adversely affects human health and causes reproductive dysfunction in laboratory animals when given at high concentrations. There is no direct evidence for adverse effects on human health at the concentrations found in contaminated groundwater. Although not really an agricultural activity, suburban lawn care can cause groundwater contamination.

Mining

The formation of acid mine-drainage water has been described above. In areas where coal mining has been carried out over long periods of time, in Appalachia and Pennsylvania, for example, it is estimated that due to its low pH, the groundwater would be unusable for decades after the cessation of mining activities. Metal-mine leachate has increased manganese in wells in Washington, caused arsenic poisoning in cattle in Idaho, and increased the radioactivity of groundwater in Wyoming.

Highway Deicing Salts

Many of the northern states use large quantities of salt in combination with abrasives such as sand on icy and snowy highways. The salt-solution runoff can percolate through adjacent soils or otherwise find its way into the groundwater via storm drains. Due to its low cost, salt is often stored in uncovered piles. Precipitation dissolves the stored salt, which may then infiltrate shallow aquifers. Large amounts of salt are used on northern highways each winter. Over 12 million tons were used in the winter of 1988-89. During 1982 and 1983 at least 9.35 million tons of dry salts and abrasives and an additional 1.78 million gallons of liquid salts were applied to highways. During a typical winter approximately 17.6 tons of salts are used per lane per mile. The amount of salt used depends, of course, upon the severity of the winter, upon the amount of abrasives used, and upon driver education. Nevertheless, salt usage in the Northeast has resulted in chloride contamination of many drinking-water wells adjacent to highways.

The dangers to public health of high salt concentrations in drinking water are not well known. Water is considered contaminated with chloride when the concentration of chloride exceeds 250 mg/liter; concentrations above that level leave a salty taste in the water. High concentrations of sodium have been linked with hypertension and have been implicated in the exacerbation of certain liver and kidney problems. The American Heart Association has suggested a maximum level of 20 mg/liter in drinking water. High concentrations of salt can also be severely detrimental to plant life.

Atmospheric Contaminants and Acid Rain

Very little is known about the effects of atmospheric contaminants and acid rain on the quality of groundwater. In Michigan chromium in the dust from an industrial plant was leached into an aquifer and contaminated a well. Hubert and Canter have reviewed the literature concerning the relation between acid rain and groundwater, and they conclude that acid rain may react with soils and surface waters in such a way as to increase the leaching of metals and nutrients. Thus, the potential for groundwater contamination is present and should be investigated further, especially the potential for lead and cadmium contamination via leaching because of their long environmentalresidence time.

Infiltration of Surface Water

Aquifers are often in hydraulic connection with bodies of surface water. Under certain conditions, polluted surface water from lakes or streams can percolate into a water-table aquifer, thereby degrading the quality of the groundwater. Also, groundwater development near a body of surface water may draw contaminated water into the aquifer.

Development of Groundwater

Pumping of water wells may bring water of lesser quality into their zone of influence. In inland areas serious problems may be caused by the migration of water from saline aquifers through leaky aquitards into potable supplies. In coastal regions extensive groundwater development has led to saltwater intrusion. Saltwater encroachment is an important source of groundwater contamination in many populated coastal regions, particularly in parts of California, Texas, Louisiana, Delaware, Florida, and New York. Saltwater occurs naturally in the aquifers in coastal areas, and pumping fresh water from wells may cause saltwater intrusion into potable aquifers, rendering them unfit for use. In 1981 the USGS estimated that 21 billion gallons of water per day, or 26% of all groundwater withdrawn, is in excess of natural recharge capabilities. In 1985 the EPA reported 19 states having problems with saltwater intrusion from deep saline aquifers or from the ocean. Attempts to prevent such contamination include the injection of freshwater "barriers" to reverse the hydraulic gradient in the aquifer so that the flow is toward the sea rather than toward the pumping wells. This method has met with some success in the Los Angeles area. Land subsidence has occurred in some areas of the United States because of excess withdrawal of groundwaters. A dramatic example of this occurred over a period of 37 years in San Joaquin Valley, California, where an area of land subsided nearly 29 feet.

Improper Construction and Maintenance of Wells

Faulty, corroded, or ruptured well casings, the linking of two aquifers by a well screen, or open wells can cause contaminants to move from one aquifer to another. Sources of contamination from the three main categories described above have been documented from all parts of the country. Whether they cause localized problems or whether the groundwater-contamination problems fall into larger regional patterns has been a matter of dispute. Our conclusions, based on a study of the documented cases of groundwater contamination available together with examples of known occurrences of groundwater contamination for selected states. It may be said, however, that groundwater contamination is not usually direct, and in most cases the contaminant must pass through the soil layer. Attenuation of contaminants has been thought to be greater in the unsaturated zone because of the greater possibility of aerobic degradation, adsorption, complexing, and ion exchange of organics and inorganics. Recent investigations on anaerobic metabolism, however, have discovered that many polluting compounds can be significantly degraded under reduced conditions. Dilution, buffering, precipitation, reduction or oxidation, mechanical filtration, volatilization, sorption, and other processes may operate within the aquifer to purify chemical wastes.

❑❑❑

9

Effects of Groundwater Contamination on Public Health

Introduction

The effects of groundwater contamination on public health may fall into one of three categories: acute waterborne disease due to a variety of pathogenic organisms, poisoning or other acute responses to chemical contaminants, and chronic long-term disease. These effects are due to three distinct forms of groundwater contamination, namely, microbial, chemical, and radioactive. Acute illnesses are defined as those resulting from contact with high levels of a chemical over a short period of time. Chronic illnesses are defined as those that persist, either because the injury is persistent or progressive, or because the exposure is prolonged and the rate of new injury exceeds the rate of repair.

Human health can be affected by groundwater contaminants through the following routes of intake: by ingestion through drinking; by inhalation of contaminants that can volatilize when water is heated (as in a shower) or boiled; by absorption through the skin during washing, bathing, or handling of soil exposed to contaminated groundwater; and by ingestion through consumption of food derived from plants or animals exposed to groundwater.

Every incidence of illness given in this chapter is attributable to contaminated groundwater. Some of the chemical causes of illness due to drinking water have been investigated for surface waters only. However, as groundwater may also contain some of these chemicals, the findings for surface waters are discussed where deemed appropriate.

Bunch and Jacobs estimated that in 1972 health hazards due to water pollution accounted for $4.4 billion of the total $188 billion estimated costs of all health hazards, but did not subdivide this total into separate costs attributable to groundwater and surface-water contamination. These figures are somewhat out of date, but more recent ones are not available.

Risk Assessment and Toxicological Testing

The degree of understanding of health effects varies among the different types of contaminants. The effects of microorganisms and high to moderately high levels of radionuclides on public health have been well studied and are now fairly well understood. The effects of very low levels of radioactivity are not so well known, and many chemical effects are not understood at all. This lack of knowledge is due to the fact that only recently has the public become aware of exposure to this type of contamination.

Table 9.1. Economic Costs of Environmental Health Hazards

Category of Hazard	Annual Health Costs Estimating (Billions of Dollars)
A. Health Costs Due to Environmental Hazards	
I. Direct Environmental Hazards	
Air pollution	8.6
Water pollution	4.4
Infectious diseases	
Hepatitis	0.73
Salmonellosis	0.0004
Influenza	4.9
Syphilis	0.007
Tuberculosis	0.7
Occupational hazards	4.4

Category of Hazard	Annual Health Costs Estimating (Billions of Dollars)
Unsafe products	6.3
Motor vehicle accidents	10.1
Total All Direct (Total I)	40.1
II. Costs Resulting from Environmentally Induced Behavior	
Alcohol abuse	24.2
Drug abuse — heroin	3.7
Tobacco use	7.6
Total All Costs of Environmental Hazards (I and II)	75.6
B. Costs of All Illnesses	188.

Assessing the risks of exposure to contaminants is, more often than not, very hard to do. The OTA points out that risk assessments require information about adverse effects, toxicity, and exposure. Epidemiological studies should be conducted when possible as well. Often, sufficient data for an assessment is not available, and especially not easily linked to the substances found in groundwater.

Toxicological testing falls into two main categories in vivo and in vitro methods. In vivo testing involves the use of animals, can be of long or short duration, and is useful for determining either chronic or acute health effects. Most toxicological data is aquired by this type of testing. In vitro testing involves the use of isolated cells or microorganisms, usually in a culture, and the testing is usually of short duration. Epidemiological studies are of three kinds: the collection and evaluation of data resulting from observations on human populations followed over time (prospective studies), the comparison of death certificates of a control population (retrospective studies), and the study of the past history of patients (case-control studies). Hubert and Canter note that epidemiological studies often produce inconclusive results because of the complexity of environmental factors that may be involved but not included in the studies. Most epidemiological evidence linking drinking-water contaminants with health effects has been based on surface-water sources.

Although the risk assessment of the ingestion of surface water cannot be directly applied to the ingestion of groundwater, because the two sources may contain quite different compounds, it nevertheless provides useful insights into the risks associated with some chemicals or classes of chemicals common to both. Absolute risks are very difficult to measure, usually apply only to the specific population for which they are determined, and are influenced by all the other pertinent characteristics of the specific population. Relative risks, however, are more likely to apply to other populations, although they too are influenced by the characteristics of the population exposed. The federal government has provided additional guidance on selected substances through the establishment of non-enforceable health goals, called MCL goals. Standards for other chemicals that may be of concern have not been set due to lack of necessary data. The debate over whether or not there is a threshold below which no adverse response occurs complicates the task of setting standards for drinking water further. Synergistic effects of contaminants should also be taken into account when MCLs are set, because contamination of groundwater is usually from a group of chemicals. In addition to the factors listed above that hinder standard setting, the cost of determining a standard can be great. Page notes that estimates of the cost of establishing carcinogenicity in one compound can range from $500,000 to $1.25 million, and the procedure is time-consuming. Very few of the contaminants in water supplies have been so tested.

Until MCLs are established there are two possibilities for action that have been suggested. Either all potential carcinogens and toxic substances could be eliminated from the water supply in case they cause future harm, or such action could be delayed until its potential benefits are better elucidated. The long time lag between exposure to a carcinogen and the appearance of malignancy is yet another complicating issue. As it could take up to 30 years to establish MCLs for contaminants on the priority pollutant list, some prudent interim measure could be adopted if deemed necessary.

Incidence of Disease

Information concerning the effects of contaminated water on public health may be obtained from two main sources: the Health

Effects Research Laboratory of the Environmental Protection Agency (EPA) in Cincinnati, and the Centers for Disease Control (CDC) in Atlanta. Each year the EPA laboratory collects information from state watersupply agencies on the occurrence of waterborne-disease outbreaks. These data are included in the annual summaries of water-related outbreaks of disease put together by the Centers for Disease Control.

The CDC defines a waterborne-disease outbreak as "an incident in which 1) 2 or more persons experience similar illness after consumption or use of water intended for drinking, and 2) epidemiologic evidence implicates the water as the source of illness. In addition, a single case of chemical poisoning constitutes an outbreak if laboratory studies indicate that the water was contaminated by the chemical. Only outbreaks associated with water intended for drinking are included." Causes of outbreaks are subdivided by system deficiencies, namely, untreated surface water or untreated groundwater, treatment deficiencies, and distribution-system deficiencies. Waterbornedisease outbreaks are reported to CDC by state health departments on a standard reporting form. CDC believes that only a fraction of the outbreaks that actually occur are reported. The efficiency of reporting outbreaks to the health authorities varies tremendously from region to region. Once reported, the interest, diligence, and resources of the state or local health authority determine the quality of the investigation. Thus, the CDC reports cannot be considered a definitive record of the incidence of disease and etiology, a fact recognized by that agency. States with good reporting systems and a staff member interested in waterborne disease might appear to have a much higher incidence of disease than states with an equal or greater actual incidence but a poor reporting system.

The number of reported incidents of outbreaks across the country has been monitored and analyzed over the past decade by these two organizations. In the period between 1971 and 1982 there were 399 outbreaks of acute waterborne disease from both surface water and groundwater reported to the CDC and the EPA; these outbreaks were caused by all types of agents and affected nearly 86,100 persons. Of these, 204 of the outbreaks (51%) and 34,337 of the illnesses (40%) were caused by the use of untreated or inadequately treated groundwater. Most of the outbreaks occurred in non-community water systems, but the greatest number of people affected resulted from outbreaks in inadequately disinfected community water systems. Craun observed that a large number of outbreaks were seasonal, affecting travelers, campers, visitors to recreational areas, and restaurant patrons during the summer months.

Of the 8558 illnesses caused by microbial contamination by the use of untreated well water, overflow or seepage through substrate of sewage from septic tanks and cesspools was responsible for the greatest number.

Table 9.2. Waterborne Disease Outbreaks in Groundwater Systems

	Outbreaks		Cases of Illness	
Water System	**Wells**	**Springs**	**Wells**	**Springs**
1. Untreated Groundwater				
A. Community	15	5	1,334	123
B. Non-community	73	8	7,029	1,116
C. Individual	22	2	195	29
Subtotal	110	15	8,558	1,268
2. Inadequately Disinfected Groundwater				
A. Community	26	4	17,125	1,519
B. Non-community	41	8	4,363	1,504
C. Individual	0	0	0	0
Subtotal	67	12	21,488	3,023
Totals	177	27	30,046	4,291

Chemical contamination was responsible for only 157 of the illnesses. Surface runoff, flooding, and improper well construction caused another 11%. Thirty-five percent of the sources of outbreaks from untreated well water have remained unidentified, Of particular interest is that most of these acute illnesses were from organisms and not from chemical poisonings. It should be remembered, however, that these reported outbreaks represent only a fraction of the total actual occurrences.

Craun suggests that the apparent trend of increased outbreaks between the years 1981 and 1990 was most probably due to improved data collection and reporting systems. Evidence suggests, however, that in recent years increased surveillance of outbreaks and their origins has led to corrective measures, and thus the numbers of reported incidents are now decreasing. This apparent decrease, however, may also be a reflection of a decrease in the numbers of people obtaining water from untreated sources.

Table 9.3. Waterborne Disease Outbreaks from Surface Water and Groundwater

Year	Number of Outbreaks	Number of Illnesses
1971	20	5,182
1972	30	1,650
1973	25	1,762
1974	25	8,356
1975	24	10,879
1976	35	5,068
1977	34	3,860
1978	32	11,435
1979	42	9,750
1980	53	20,045
1981	35	4,450
1982	44	3,611
Totals	399	86,048

Table 9.4 lists some of the disease-causing pathogens and chemicals that are known to be responsible for outbreaks in the United States, as well as the clinical symptoms or laboratory or epidemiological criteria required for their confirmation.

Table 9.4. Summary of Toxic Effects of Known Groundwater Contaminants

Key: The numbers after each chemical correspond to the numbers that precede each toxic effect listed below.

1. Eye irritation
2. Skin irritation
3. Allergic sensitization
4. Upper respiratory tract irritation
5. Lung/respiratory effects
6. Liver damage
7. Kidney damage
8. Pancreatic damage
9. CNS effects
10. Peripheral nervous system effect

Table Contd.

11.	Blood cell disorders	18.	Endocrine effects
12.	Immunological effects	19.	Reproductive effects
13.	Cardiovascular effects	20.	Embryotoxicity
14.	Gastrointestinal effects	21.	Teratogenicity
15.	Cholinesterase inhibition	22.	Mutagenicity
16.	Methemoglobinemia 127	23.	Carcinogenicity
17.	Skin damage		

I.	Organic Chemicals Other than Pesticides	Toxic Effects
A.	Aliphatic Hydrocarbons	
	Cyclohexane	6, 7, 9
	Propane	13
	Dicyclopentadiene	1, 2, 4, 5, 7, 9
	(Note: many petroleum products are in this class; see III below)	
B.	Aromatic Hydrocarbons	
	Benzene	1, 2, 3, 9, 10, 11, 12, 14, 24
	Naphthalene	1, 2, 4, 11, 18
C.	Polynuclear Aromatic Hydrocarbons	
	Acenaphthalene	2, 4
	Anthracene	3, 23, 24
	Chrysene	24
	Dibenz[a.h.]anthracene	24
	Phenanthracene	3, 23, 24
	Pyrene	2, 5, 6, 11, 14, 22, 23
D.	Alkyl-Substituted Aromatic Hydrocarbons	
	Ethylbenzene	1, 2, 4, 6, 7, 9
	n-Propylbenzene	10
	Styrene (unsaturated side chain)	1, 2, 4, 6, 7, 9, 14, 21, 23

Table 9.5 lists toxic substances that have been found in groundwater, and their effects; radioactive substances, however, are not included.

Acute Illness Due to Contaminated Groundwater

Fortunately, most acute waterborne poisonings cause mild, selflimited illnesses. The symptoms vary with each substance ingested. However, commonly reported symptoms suggestive of acute illness include gastrointestinitis, nausea, vomiting, and skin rashes; these effects usually occur soon after exposure.

Acute Illness Due to Pathogenic Contaminants

Five categories of pathogens are found in water: bacteria, viruses, protozoa, worms, and fungi. The most significant waterborne diseases worldwide are typhoid and cholera, which were a scourge in the industrialized nations, including the United States, up to the late nineteenth century. Gastrointestinal illnesses constitute the major category of illnesses resulting from groundwater contamination.

Bacteria

Of the more than 400 types of bacteria present in human and animal feces, Escherichia coli, an organism that is not usually pathogenic and is always found in the gastrointestinal tracts of humans and other warm-blooded animals, has been used for most of this century as a monitor of the safety of water for human consumption. The potential for human infection by pathogenic bacteria depends upon the bacterial survival rate and the contamination potential, which, in turn, is affected by the type of soil, the rate and duration of bacterial loading, the groundwater flow, and the integrity of the media surrounding the aquifer. Survival times may vary from less than one day for *Salmonella typhi, Vibrio cholerae,* and *E ntamoeba histolytica* cysts on vegetables to several months for *Salmonellae* in soil and *Shigellae* in water containing humus. The viability of bacteria in porous media under favorable conditions in the tropics may be five years, although 60 to 100 days might be the maximum in temperate regions.

Table 9.5. Summary of Toxic Effects of Known Groundwater Contaminants

I. Organic Chemicals Other than Pesticides	Toxic Effects
Toluene	1, 2, 4, 6, 7, 9, 13
1,2,4-Trimethylbenzene	2, 5, 9
Xylenes	1, 4, 6, 7, 9, 13

Table Contd.

E. Aliphatic Alcohols	
Benzyl alcohol	2
128 Methanol	1, 6, 9, 18
F. Phenols	
p-Tert-butylphenol	2
G. Ethers	
Tetrahydrofuran	2, 4, 6, 7, 9
Dioxane	1, 4, 6, 7, 24
Diethyl ether	1, 2, 4, 9
H. Ketones	
Acetone	1, 4, 9
Benzophenone	3
I. Carboxylic Acids	
Acetic acid	1, 2, 4, 5
Formic acid	1, 2, 4, 9
J. Halogenated Aliphatic Hydrocarbons	
Bromochloromethane	5, 6, 7, 9
Bromoform	1, 4, 6, 9
Carbon tetrachloride	6, 7, 9, 20, 24
Chloroform	1, 2, 6, 7, 9, 14, 21, 24
Chloromethane	5, 6, 7, 9
Dibromochloromethane	23
1,1-Dichloroethane	6, 7, 9, 21
1,2-Dichloroethane	1, 2, 5, 6, 7, 9, 13, 14, 19, 20, 23, 24
1,1-Dichloroethylene	1, 2, 4, 5, 6, 7, 9, 13, 19, 21, 23, 24
Dichloromethane	1, 2, 6, 7, 9, 21
1,2-Dichloropropane	6, 7, 23
Fluoroform	23
Hexachlorocyclopentadiene	5, 6, 7, 13, 14, 19
Hexachloroethane	6, 7, 9, 21, 24
1,1,1,2-Tetrachloroethane	6, 21, 23
1,1,2,2,-Tetrachloroethane	4, 6, 7, 9, 14, 21, 22, 23 24
Tetrachloroethylene	1, 4, 5, 7, 9, 21, 24
1,1,1-Trichloroethane	1, 2, 5, 6, 7, 9, 13, 23, 24

1,1,2-Trichloroethane	1, 2, 4, 5, 6, 7, 9, 23, 24
Trichloroethylene	2, 5, 6, 7, 9, 11, 13, 24
Trichlorofluoromethane	9, 13
Vinyl chloride	2, 6, 7, 9, 11, 23, 24
K. Halogenated Aromatic Hydrocarbons	129
Bromobenzene	1, 2, 5, 6, 8, 23
Chlorobenzene	1, 4, 5, 6, 7, 9, 20, 23
2-Chloro-naphthalene	2, 6
o-Chlorotoluene	1, 2, 9
o-Dichlorobenzene	1, 2, 4, 5, 6, 7, 9, 11
p-Dichlorobenzene	1, 4, 5, 6, 7, 9, 11
Polybrominated biphenyls	6, 12, 21, 22
Polychlorinated biphenyls	5, 6, 12, 13, 14, 20, 21, 22, 23, 24
L. Aromatic Amines	
Aniline	9, 16, 24
Benzidine	24
M. Esters	
Bis-(2-ethylhexyl)phthalate	1, 2, 6, 20, 21, 22, 24
Dibutyl phthalate	1, 4, 9
Diethyl phthalate	2, 4
N. Chlorophenols	
Pentachlorophenol	1, 2, 4, 6, 9, 21
2,4,6-Trichlorophenol	6, 23, 24
O. Multifunctional and Miscellaneous Structural Types	
Alkyl benzene sulfonates	1, 2, 3, 14
Alkyl sulfonates	1, 2, 3
Benzoyl chloride	2, 4
Chlorinated dibenzo-p-dioxins	6, 9, 10, 17, 20, 22, 23, 24
4,4'-Methylene-bis-2-chloroaniline (MOCA)	1, 4, 6, 7, 16, 23, 24
Dichlorobenzidine	1, 4, 5, 14, 24
Dimethyl disulfide	22
4-Dinitrosodiphenylamine	24
Dodecylmercaptan	1, 2, 3, 9, 23
o-Nitroaniline	16, 23

Table Contd.

Nitrobenzene	6, 7, 9, 11, 16, 19, 20, 21
4-Nitrophenol	5, 16
N-Nitrosodiphenylamine	6, 24
II. Pesticides	Toxic Effects
A. Chlorinated Hydrocarbons	
Aldrin	2, 6, 7, 9, 20, 24
1,2-Dibromo-3-chloropropane 130 (DBCP)	1, 2, 5, 6, 7, 20, 23, 24
Chlordane	4, 6, 9, 13, 24
DDT	5, 6, 9, 11, 20, 23, 24
DDD	19, 24
Dieldrin	6, 9, 20, 24
Endosulfan	6, 7, 9, 21, 23
Endrin	6, 9
Kepone	6, 7, 20, 21, 24
Methoxychlor	7, 20
Toxaphene	6, 7, 9, 21, 23, 24
B. Carbamates	
Aldicarb	15
Carbofuran	6, 15
C. Organic Phosphates	
Chlorpyrifos	6, 15
Dichlofenthion	15
Parathion	15, 20, 21
Phorate	15
D. Phenoxy Herbicides	
2,4-Dichlorophenoxyacetic acid (2,4-D)	9, 10, 22, 23
2,4,5-Trichlorophenoxyacetic acid (2,4,5-T)	6, 7, 9, 22, 23
Silvex	6, 7, 9
E. Other Herbicides	
Atrazine	2, 5, 9, 11, 21, 23
Bromacil	6, 9
Simazine	9, 23
F. Miscellaneous	
2,4-Dinitrophenol	1, 2, 7, 9, 20

III. Petroleum Mixtures	Toxic Effects
Gasolines	2, 4, 9, 14
Fuel oils and diesel	2, 5, 7, 9, 14
Jet fuels	6, 7, 9, 10
IV. Inorganic Chemicals	Toxic Effects
Ammonia	1, 2, 4, 5, 6, 7, 9
Cyanide	1, 2, 3, 4, 9
Fluoride	7, 9
Aluminum	14
Antimony	2, 5, 6, 7, 11, 13
Arsenic	5, 10, 13, 17, 20, 22, 23, 24
Barium	2, 4, 5, 7, 10, 19
Beryllium	1, 2, 3, 4, 5, 6, 24
Cadmium	4, 5, 7, 9, 20, 21, 22, 24
Chromium	2, 3, 4, 5, 6, 21, 22, 23, 24
Cobalt	1, 2, 3, 5, 11, 13, 14
Copper	6, 7, 11
Iron	5, 6, 14
Lead	9, 11, 20
Lithium	7, 9, 13, 14
Magnesium	1, 9
Manganese	5, 6, 9
Mercury	7, 9, 18, 22
Nickel	1, 3, 4, 5, 9, 14, 20, 24
Palladium	7, 11
Selenium	1, 4, 5, 6, 9, 14
Silver	5, 7, 11, 14, 17, 18
Thallium	6, 7, 9, 13, 14, 20, 22
Vanadium	3, 5, 6, 7, 9, 11
Zinc	1, 2, 11, 19, 20

Bacteria in water flowing through porous media are removed by filtration, die-off, and adsorption. In medium-grained sand and finer materials, pathogenic coliform organisms only penetrate a few yards, whereas in sand or gravel aquifers such transport may be in the order of hundreds of yards. Transport in fractured rock, where relatively high groundwater velocities occur (upwards of several feet per day), 132 permits transport in the order of miles for a microorganism whose viability may range from days to months.

EPA-approved testing for bacterial contamination is standardized throughout the United States and consists of three phases: a presumptive test, a confirmed test, and a completed test. These procedures are summarized by NAS. The information provided by the EPA laboratory in Cincinnati and by the CDC annual summaries shows that between 1945 and 1984 nationwide there were 104 groundwater-related outbreaks of disease of bacterial origin causing 26,763 cases of illness.

Viruses

More than 100 different enteric viruses have been identified in human fecal material, and these may survive for several months in wastewater, tapwater, soil, and shellfish. In addition, the normal practice of wastewater chlorination may not be sufficiently effective to inactivate viruses. The diseases caused by enteric viruses range from rashes, diarrhea, and fever to meningitis, paralysis, encephalitis and myocarditis and are summarized in Table 9.6.

The sources of infection include contaminated water and food pathways. The minimum infective dose of ingested viruses depends on the type of virus involved, the route of entry, and the susceptibility of the host. Doses as low as a single infectious unit may cause disease in man. WHO report concludes that current epidemiological techniques are not sensitive enough to detect low-level transmission of viral diseases through water for two reasons. Firstly, most enteric viruses cause such a broad spectrum of symptoms that the identification of the single entiological agent causing the illness is often too difficult to identify. Secondly, viral agents causing infections are difficult to recognize as being waterborne. Nearly 60% of the recently reported cases of waterborne disease in the United States were caused by unknown or unrecognized agents. Infectious hepatitis can often be attributed to the waterborne category.

The principal sources of viral contamination of groundwater contain human and animal fecal material. They include sewage effluent disposed of in injection wells, land-waste disposal, animal 134 feedlots and dairies, septic tanks, and cesspools.

Table 9.6. Human Enteric Viruses That May Be Present in Water

Virus Group	No. of Types	Disease Caused
Enteroviruses:		
Poliovirus	3	Paralysis, meningitis, fever
Echovirus	34	Meningitis, respiratory disease, rash, diarrhea, fever
Coxsackievirus A	24	Herpangina, respiratory disease, meningitis, fever
Coxsackievirus B	6	Myocarditis, congenital heart anomalies, rash, fever, meningitis, respiratory disease, pleurodynia
New enteroviruses	4	Meningitis, encephalitis, respiratory disease, acute hemorrhagic conjunctivitis, fever
Hepatitis type A (probably an enterovirus)	1	Infectious hepatitis
Gastroenteritis virus (Norwalk type agents)	2	Epidemic vomiting and diarrhea, fever
Rotavirus (Reoviridae family)	?	Epidemic vomiting and diarrhea, chiefly of children
Reovirus	3	Not clearly established
Adenovirus	30	Respiratory disease, eye infections
Parvovirus (adeno-associated virus)	3	Associated with respiratory disease in children, but etiology not clearly established

Land application of wastewater for irrigation and for disposal may also lead to viral contamination of the groundwater. The movement of viruses in the soil depends upon the rate of application, the composition and structure of the soil, its pH, organic content, and strength of the effluent. Viruses are readily absorbed onto clays under suitable conditions and are also removed by sandy loams and soils with a high organic content. Fractured rock permits the viruses to travel great distances unimpeded.

Current methods of monitoring viral contamination are summarized in Table 9.7 and require special facilities: thus they are not routinely performed on drinking-water samples. Sewage-treatment processes are not likely to remove all viruses present. Sedimentation may remove 50%; the activated-sludge process, 60%-99%; trickling filters and stabilization ponds, 80%-95%. Chemical coagulation with lime achieves a 90%-99% reduction.

Information from the EPA laboratory in Cincinnati and the CDC's annual reports show that nationwide between 1955 and 1994 there were 64 outbreaks of disease due to groundwater-related viruses. These resulted in 4,365 cases of illness. The majority of the outbreaks were due to the hepatitis virus, which caused 51 outbreaks and affected 1,919 people. The total number includes 9 outbreaks of the Norwalk-type viral agents, affecting 1,476 persons; 3 outbreaks due to parvovirus, affecting 937 people; and 1 outbreak of polio virus affecting 16 persons.

Parasites

This category covers protozoa, worms, and fungi. Hubert and Canter report no recorded incidents of the outbreak of disease attributable to the contamination of groundwater by a parasitic organism. Outbreaks due to contamination of surface water by *Giardia, Entamoeba histolytica,* and *Ascaris lumbricoides* were reported, and Hubert and Canter concluded that the absence of outbreaks due to the contamination of groundwater is probably attributable to the size of the organisms involved.

The EPA, however, in a printout of the incidence of disease due to groundwater contamination from 1955 to 1990, reported that 9 outbreaks of protozoan parasitic disease were reported between 1963 and 1990. An additional 9 outbreaks were reported by the CDC between 1991 and 1994. These 18 included major outbreaks of illness due to *Giardia* in Arizona, Colorado, Florida, Virginia, Wisconsin, Vermont, and Pennsylvania, and to *Entamoeba histolytica* in Arizona, Indiana, Missouri, Mississippi, Oklahoma, and South Carolina.

Giardia lamblia is the most commonly reported protozoan pathogen that causes outbreaks of disease from water supplies. Between 1975 and 1992 there were 54 outbreaks caused by *Giardia,* affecting approximately 20,000 people. However, one must remember

Table 9.7. Methods for the Detection of Viruses In Different Types of Water and Sludges

Method	Type of Sample
Direct inoculation without concentration	Wastewater, sludge
Swab sampling	Wastewater samples (method sensitive but not quantitative)
Filter adsorption elution methods	0.2-5.0-liter samples where more than 1 infectious unit of virus per liter is expected
Adsorption and precipitation methods employing polyvalent cation salts	
Use of preformed alum, aluminum hydroxide, iron (111) oxide (50), iron (111) chloride, or lime flocs	
Flocculation by added salts	
Hydroextraction and aqueous polymer two-phase separation techniques	
Soluble alginate	
Precipitation by low pH	
Flat membranes, hollow fibre	Large-volume samples, of 5-400 liters or more. In general, single- stage procedures may be used for samples of up to 20 liters i.e., for samples containing relatively large amounts of virus such as are likely to occur in sewage, treated sewage effluents, and polluted surface water
Flow-through filter adsorption of acidified samples followed by elution at high pH	
Single or multistage procedures	
Filtration through pleated filters followed by elution	Groundwaters, less polluted surface waters, and highly treated wastewater (multistage elution procedures)
Organic flocculation followed by elution	
Filtration through adsorptive filters with positive charge followed by elution	
Adsorption to glass powder followed by elution	

that in only about one-half of all of the cases reported is the cause of disease ever established; therefore, both of these numbers may be gross underestimates. Of the *Giardia* outbreaks reported, 72% of the cases involved contaminated surface water, 13% involved groundwater, and in 15% of the cases the water source was not reported. Most of the cases were a result of untreated drinkingwater supplies or deficiencies in the distribution or treatment processes.

Giardia passes through four stages in its life cycle. The third stage, the cyst, is the most infective. In this stage the parasite has great resistivity to chlorine and can survive in water for more than two months. The virus can be contracted through contact with an infected person or animal or through consumption of water contaminated with the pathogen. Although boiling destroys the cyst, the most effective treatment appears to be chemical pretreatment (coagulation) followed by filtration and chlorination.

Recently, federal regulation for controlling the incidence has been established under the Safe Drinking Water Act Amendments. An MCL goal of 0 for Giardia was proposed by EPA and is now included in the list of 83 contaminants being considered for MCLs. An EPA-sponsored conference on the Microbial Health Considerations of Soil Disposal of Domestic Wastewaters was held by the National Center for Ground Water Research (NCGWR) in 1981. The aim of the conference was to identify important health related research needs associated with the deposit of domestic wastes on land via septic tanks, cesspools, and land-application systems. The papers included reviews of the current practices of disposal, the transport and fate of various classes of pathogens, outbreaks of waterborne disease, risk evaluations, and possible control measures. Pathogen removal by soils in the land application of wastes can be improved by changing the infiltration rate. Increased pollution is often traced to the contaminated recharge water from partially treated wastes that comes in contact with groundwater. Sobsey concludes that the information on the transport and fate of viruses is inadequate, especially on the mechanisms of microbially mediated antiviral activity in soils. Gerba states that the current body of information concerning viruses in groundwater is not sufficient to develop guidelines for estimating safe distances between waste sources and drinkingwater wells and proposes that this gap be filled by further field studies. Cabelli describes the criticism levelled at the microbial-indicator systems most frequently used to monitor drinking water in the United States, the total-coliforms and the standard-plate count. It is thought that coliform organisms do not survive environmental stress, including disinfection or residence in wastewaters, as well as do some of the viral enteric pathogens. He suggests that alternative indicators, such as enterococci, *Clostridium perfringens* spores, acid-fast bacilli, yeasts, and coliphage, would better reflect the survival characteristics of viral pathogens and noted that two of these are used along with *E. coli* as indicators in Europe. Deficiencies in the indicator system could mean that the risk of disease is understated. An evaluation of the risk to public health of various methods of disposing of wastewater explored by Cooper and Olivieri uses a probability-matrix technique that requires data from two groups of experts. The first group delineates the health problems associated with a given process and estimates the probability of their occurrence, and the second group judges the severity of the problems delineated by the first group. Boyle describes the major causes of subsurface-disposal failure leading to contamination and suggests technical and managerial remedies, particularly for septic-tank failures. Despite its aim, the conference does not appear to have produced a list, in order of priority, of the health-related research needs.

When groundwater is used for drinking water by municipal suppliers, pretreatment, such as chlorination, minimizes the possibility of outbreaks of disease caused by pathogens. Deficiencies in treatment cause problems. In domestic wells with no pretreatment of the water, prevention of contamination is of paramount importance.

Acute Illness Due to the Chemical Content of Water

Acute illness due to chemical poisoning is reported in the EPA computer printout of the incidence of illness attributable to the ingestion of contaminated groundwater. Hubert and Canter listed both the known cases of chemical contamination affecting public health and their accompanying symptoms. In Craun's *Summary of Waterborne Illness Transmitted through Contaminated Groundwater*, 45 (11%) of the 399 outbreaks that occurred between 1971 and 1980 were due to chemical poisonings. These chemicals, natural and man-made, organic and inorganic, included petroleum products (lubricating oil, fuel oil, gasoline), nitrate, polychlorinated biphenyls, trichloroethylene, phenols, benzene, arsenic, and selenium. Only 157 persons were affected by these outbreaks.

Harris reports that many of the acute water poisonings that have occurred in the past have been due to fluoride equipment failure, contaminating drinking-water sources with excess amounts of flouride, and heavy metal poisoning, occurring mainly as a result of corrosion of plumbing materials by acidic waters. Organic chemical poisonings have resulted from many factors, but mainly from accidental spills.

Recent investigations have shown widespread contamination of groundwater by toxic organic chemicals. In 1984 the National Academy of Sciences (NAS) reported that there is virtually no toxicity information in existence for approximately 70% of the estimated 66,000 chemicals currently in commercial use. The NAS itself has studied approximately 21 organic chemicals now found in drinking water.

Estimates for the number of chemicals in existence today vary greatly. Weimar noted that approximately 1,000 new chemicals enter the environment each year, in addition to an estimated 30,000 that already exist there. Estimates range as high as 75,000, but not all of these chemicals would find their way into landfills or impoundments, nor would all of them prove harmful to human health.

Many of the organic chemicals recently found in groundwater originate from industrial waste. However, a common inorganic contaminant, nitrate, comes mainly from natural sources, sewage disposal, animal feedlots, and the use of agricultural fertilizer. The Maximum Contaminant Level (MCL) established by EPA for nitrate was set at a level to prevent infantile methemoglobinemia. This disease occurs almost exclusively in infants less than three months old who have ingested high levels of nitrate. Before or after ingestion, nitrate is bacterially converted to nitrite. In the body this nitrite oxidizes the ferrous iron in hemoglobin to ferric iron, converting hemoglobin to methemoglobin. Methemoglobin is incapable of binding molecular oxygen, and thus cyanosis, respiratory distress, or suffocation of the infant may result. Death may occur from the ingestion of water containing 50 100 mg/l NO3. Cases of blue-baby syndrome have not been well documented, and national statistics are not available. Hallberg, however, reports that four acute cases have been reported in western Iowa, eastern Nebraska, and southeastern South Dakota in recent years. Only one incident of death due to high nitrate levels from public water supplies has been reported since 1960, and few cases due to concentrations from individual supplies have been reported in the past 10 years. A chronic health hazard associated with the ingestion of nitrites is their interaction with amines or amides within the body to form nitroso compounds, nitrosamines, and nitrosamides. These substances have been shown to produce cancer in laboratory animals.

Local deposits of arsenic, selenium, and other naturally occurring inorganic compounds may contaminate groundwater and render it unfit for human consumption. Selenium and other chemicals, which are toxic in high concentrations and even carcinogenic, are also required nutrients. In general, very little is known about the acute effects caused by the ingestion of the chemicals listed above; however, many of the effects they produce through inhalation and skin absorption have been summarized. These effects, however, may differ somewhat from the effects caused by ingestion. In conclusion, it may be said that relatively few cases of acute effects on public health have been documented in response to chemical contamination. This is in part caused by our lack of knowledge of the effects on human health of many chemicals.

Chronic Effects of Contaminated Groundwater on Public Health

Incidents suggestive of the chronic effects on public health of groundwater contamination have been reported for cancer, malformations, miscarriage, central nervous system disorders, and cardiovascular disease, although there have been few carefully controlled epidemiologic investigations that confirm these associations. The controversy surrounding the question of whether contamination can have one or a variety of long-term effects may be divided into two parts, namely, the effects associated with the mineral characteristics of the water supply and those that may be associated with its organic chemical contamination. Each will be dealt with in turn.

Chronic Disease Due to the Mineral Content of Groundwater

Early studies of the relation between the mineral content of water and apoplexy in Japan and cardiovascular disease in the United States raised the possibility that a so-called water factor was involved believe that this controversy will continue even though the American Medical Association has stated that adverse effects on health are probably due to life-style rather than to environmental factors. Hubert and Canter give an extensive summary of all the literature available on the relation between the hardness of water and coronary disease, arteriosclerosis, ischemic heart disease, hypertension, and stroke. In addition, they review the inherent weaknesses of the studies to date. Hubert and Canter summarize the conclusions to be drawn from the available literature as follows:

- The relation between water hardness and cardiovascular disease and mortality is uncertain.
- Cardiovascular disease and mortality is known to be greatly influenced by factors varying from diet to stress, and the situation is further complicated by the mineral content of potable groundwater.
- Although many studies show a statistical correlation between water hardness and cardiovascular disease, the responsible element or compound in water has not been conclusively identified.

Because of the uncertainty surrounding the water factor, Hubert and Canter conclude that it is impossible to design a water treatment for potable groundwater to remove allegedly harmful 141 constituents related to cardiovascular disease.The effects on health of inorganic solutes, organic solutes, and radioactive substances in drinking water have been admirably summarized in the study on drinking water and health carried out by the National Research Council of the National Academy of Sciences. In a later continuation of this study the epidemiological studies of water hardness and cardiovascular disease were also assessed. The panel noted that current knowledge is derived mainly from ecological epidemiological studies, those in which individual exposures and risk factors were not considered. A prior study, *The Geochemistry of Water in Relation to Cardiovascular Disease*, came to the following conclusions:

- Studies encompassing large geographical areas generally showed a correlation between hard water and low cardiovascular-disease rates. This correlation breaks down for smaller study areas or if the population studies are grouped by altitude or by proximity to a seacoast. The association of some noncardiovascular illnesses with soft-water areas raises the possibility that soft water may be a concomitant of some more basic risk factor.
- Most studies reported correlation coefficients rather than risk estimates as a function of exposure. Upper estimates of risk ratios for soft compared with hard water, from those studies with sufficient data to be useful in such an assessment, average approximately 1 to 25 for cardiovascular disease and 1 to 2 for arteriosclerotic, stroke, and hypertensive disease.
- Autopsy studies have reported low magnesium levels in the tissue of the heart, diaphragm, and pectoral muscle of persons who die from myocardial infarction as compared with those who die from accidental causes. Similar deficits were observed in persons living in soft-water areas as compared with hard-water areas. These findings are in line with those of Sharrett, Comstock and Hubert and Canter.

It is recognized that while certain trace metals are essential to human nutrition, higher concentrations may be harmful. Trace metals may be introduced into groundwater as a result of the natural local occurrence of specific minerals in the aquifer or the disposal of certain types of industrial waste. Berg and Burbank point out that present-day occurrences of cancer should be 142 correlated with trace-metal levels of 20 to 30 years ago, but such data are not available. They compared mortalities for 34 types of cancer with the incidence of eight metals in surface water in the United States. The positive correlations of 28 were double the number expected. No significant correlations were found for iron, cobalt, or chromium. The correlations for lead were positive for kidney cancer, all lymphoma and leukemia, and intestinal cancer mortalities, which correspond to the known biological activities of lead. Beryllium was significantly correlated with all cancers, and this, the authors considered, also reflected its biological activity. Arsenic and nickel are thought to be human carcinogens. It should be mentioned that arsenic compounds have not been reliably demonstrated to produce tumors in laboratory animals, but epidemiological studies show that the incidence of certain cancers and precancerous conditions increases in humans who are chronically exposed to arsenic compounds by oral or respiratory routes. The results are shown in Table 9.8. Several milligrams per liter of arsenic have been found naturally in the groundwaters of Indiana, Utah, Alaska, Oregon, and California. Cadmium showed the strongest correlations with the most types of cancer, and although dose-response relationships are uncertain, the authors suggest further studies.

Asbestos fibers were discovered in drinking water and beverages throughout the country, resulting in an increased concern over the possible health effects associated with ingestion. Asbestos is a group of naturally occurring silicate minerals that have a fibrous morphology and are found in drinking water as a result of both natural and industrial processes. Recent studies have concluded that lung cancer and mesothelioma has occurred in experimental animals after inhalation of asbestos. However, similar effects were not found in these same animals after ingestion of the same amounts of asbestos. It has also been reported that epidemiologic studies conducted in the United States and Canada have yielded no definitive conclusions regarding adverse effects of ingested asbestos. Likewise, none of the epidemiologic studies of human exposure through ingestion have been able to define levels that pose a risk. Despite the inconsistent results of these asbestos studies, EPA is currently considering regulating asbestos levels in public drinking waters.

Table 9.8. Statistically Significant Positive Correlations Between Metal Concentrations and Cancer Death Rates

Metal	Cancer (ICD)	Probability of Positive Association
Arsenic	Larynx (161)	0.024
	Eye (192)	0.009
	Myeloid leukemia (204.1)	0.042
Beryllium	Breast (women) (170)	0.040
	Uterine cervix (171)	0.016
	Other uterus (172-174)	0.006
	Bone (196)	0.024

Metal	Cancer (ICD)	Probability of Positive Association
	All cancers (140/205)	0.016
Cadmium	Mouth and pharynx (140/149)	0.003
	Esophagus (150)	0.00004
	Large intestine (153/154)	0.00001
	Larynx (161)	04
	Lung (162/163)	0.001
	Breast (women) (170)	0.003
	Bladder (181)	0.009
	Myeloma (203)	0.008
	All lymphomas (200/203, 205)	0.016
	All cancers (140/205)	0.0005
Chromium	None	
Cobalt	None	
Iron	None	
Lead	Stomach (151)	0.0026
	Small intestine (152)	0.038
	Large intestine (153/154)	0.009
	Ovary (175)	0.02
	Kidney (180)	0.008
	Myeloma (203)	0.042
	All lymphomas (200/203, 205)	0.0005
	All leukemias (204)	0.006
Nickel	Mouth and pharynx	0.044
	Large bowel	0.031

Trace metals may cause additional chronic health problems other than cancer. The Environmental Protection Agency reported that the accumulation of lead in bone and tissue may have deleterious effects on the nervous system and kidneys, and may 144 also cause circulation problems. In some cases high doses of lead in children have been shown to affect intelligence, behavior, and school performance. Effects of mercury poisonings include central nervous system dysfunctions, eye damage, urinary tract damage, embryo injury, physiological disturbances, and even death. High doses of cadmium accumulate in the liver and kidneys and may eventually cause kidney disease in old age. There is some indication that cadmium may cause birth defects as well. Elements such as arsenic, selenium, and nitrates may also create health problems at high levels. Arsenic has been reported to damage human gastrointestinal tracts and cause heart abnormalities. Selenium poisoning may cause damage to the nervous system and the gastrointestinal tract, and high levels have been known to induce dental decay. Nitrate at high levels, as mentioned earlier, may cause methemoglobinemia and cancer. In addition, it has been reported to cause nervous system impairments and birth defects.

Some evidence indicates that low levels of certain minerals may have a protective effect against cancer and other pathologies. Pories, stating that iodine is the best known inhibitor of experimental neoplastic growth, also found that such inhibitory properties were shown by arsenic, copper, platinum, selenium, and zinc. Inclusion of arsenic in the diet decreased the incidence of all tumors in animals and also stunted carcass growth. Copper is known to potentiate the biological activity of certain anti-cancer agents. This could be due to the fact that it is a potent free radical inhibitor or that it shields specific sites on liver protein molecules. It is known that there is a high frequency of preexisting goiter, a condition caused by iodine deficiency, in patients with thyroid cancer. Platinum is known to halt cell division, and although the neoplastic inhibitory mechanism is unknown, it may be due to a platinum-induced release of viral genomes initiating an antibody reaction against the tumor. Dietary selenium of 1 ppm reduced the incidence of animal tumors. It is known that cancer death rates are lower in areas with drinking water containing high levels of selenium from natural sources. These seleniferous areas are found in Wyoming, Colorado, and South Dakota. In the past, concerns that selenium may cause cancer have been expressed; however, the NAS does not believe that scientific evidence is able to establish that selenium is indeed carcinogenic. Zinc deficiency decreases tumor growth in animals, but its role in human cancer is not well understood, and the data are contradictory. Concentrations of fluoride in drinking water between 145 0.7 and 1.2 mg/l are thought to be beneficial in preventing dental caries. However, with the ingestion of excess concentrations, mottling of teeth and dental fluorosis may result in children, and increased density and calcification of the bone may occur in adults.

Thus, although trace minerals are necessary to health, higher concentrations may be harmful. Excessive concentrations of certain minerals have been implicated in the various chronic forms of cardiovascular disease and cancer, but the studies to date are not conclusive. It is possible that water containing certain heavy metals may also protect against a broad spectrum of human tumors. Although the evidence is fragmentary, it deserves further investigation.

Chronic Disease Due to Contamination by Toxic Organic Chemicals

Burmaster's comprehensive report. *Contamination of Ground Water by Toxic Organic Chemicals*, was published by the Council on Environmental Quality. He reviewed the sources of toxic organic contamination, methods of prevention, incidents involving drinking-water contamination, and estimates of the health risks due to toxic organic chemicals. The Safe Drinking Water Committee of the National Academy of Sciences has also reviewed the effects of organic solutes on human health and the toxicity of selected organic contaminants. The Environ Corporation conducted an extensive study of the effects of toxic chemical contamination for the Office of Technology Assessment.

Many toxic organic chemicals are colorless, tasteless, and odorless in the concentrations at which they occur in drinking water. Often, chemicals found in groundwater are in concentrations orders of magnitude higher than those found in surface water. This phenomenon is a reflection of the lack of mixing and the minimal dispersion that occurs in contaminated plumes of groundwater in comparison with surface water.

Page, using data collected by the New Jersey Department of Environmental Protection, compared groundwater and surface water for patterns and levels of contamination by toxic substances. Samples from 1,000 wells and 600 surface-water 146 supplies in New Jersey were analyzed for 56 toxic substances: 27 light chlorinated hydrocarbons, 20 heavy chlorinated hydrocarbons, and 9 heavy metals. The probabilities of detecting them are shown in Table 9.9. For 64% of the toxics, the highest concentration was found in groundwater, and Page considered this a potentially significant threat to some persons consuming groundwater. Similarly high concentrations of contaminants in groundwater were also found in other states and reported to CEQ. These are shown in Table 9.10 states that in New Jersey, since groundwater is at least as contaminated as surface water and the patterns of contamination are similar for the two, the control of toxic chemicals in groundwater should receive attention equal to that given their control in surface water.

Greenberg also using data collected by the New Jersey Department of Environmental Protection, examined population exposure to toxic substances in New Jersey. The State of New Jersey organized a research program in response to the news that its residents showed the highest white male cancer rate in the United States. Non-white male and female and white female rates were also among the highest in the country. These findings have been criticized for lack of proper controls for critical variables. Analyses for 45 substances were carried out on 408 groundwater samples, representing a cross section of locations and land and water uses. The results showed that pesticide contamination tends to occur in agricultural and forest areas, and light chlorinated hydrocarbon pollution in urban areas.

The contamination of groundwater associated with hazardous chemicals has come to light only in the last decade, mainly because these chemicals have not been routinely monitored in the past. Through a number of surveys conducted in the past 10 years by the states and the federal government, several organic chemicals have been frequently detected in groundwater-supplied drinking water. The OTA and the EPA have both reported that serious contamination of groundwater by associated toxic chemicals has occurred in at least 34 states.

Table 9.9. Probabilities of Detecting Toxics in New Jersey Groundwater and Surface Water

	Probability Detectable in Groundwater	Probability Detectable in Surface Water	Significantly Different at 0.05 Level
Fluoroform	0.03	0.08	Yes
Methyl chloride	0.01	0.04	Yes
Vinyl chloride	0.01	0.03	Yes 147

	Probability Detectable in Groundwater	Probability Detectable in Surface Water	Significantly Different at 0.05 Level
Methylene chloride	0.23	0.45	Yes
Chloroform	0.64	0.64	no
1,2-Dichloroethane	0.10	0.12	no
1,1,1-Trichloroethane	0.78	0.79	no
Carbon tetrachloride	0.64	0.68	no
1,1,2-Trichloroethylene	0.58	0.56	no
Dichlorobromoethane	0.34	0.43	yes
1,1,2-Trichloroethane	0.07	0.09	no
Dibromochloromethane	0.14	0.18	yes
1,2-Dibromoethane	0.08	0.06	no
1,1,2,2-Tetrachloroethylene	0.43	0.88	yes
Bromoform	0.22	0.33	yes
1,1,2,2-Tetrachloroethane	0.06	0.11	yes
Diiodomethane	0.06	0.02	yes
Total dichlorobenzene	0.03	0.07	yes
m-Dichlorobenzene	0.02	0.04	yes
p-Dichlorobenzene	0.03	0.06	yes
o-Dichlorobenzene	0.03	0.03	no
Aroclor	1242	0.11	0.08
			no
Aroclor	1248 0.06	0.14	yes
Aroclor	1254 0.03	0.14	yes
gem-Dichloroethylene	0.44	0.65	yes
Dibromomethane	0.12	0.28	yes
trans-Dichloroethylene	0.51	0.63	yes
Bromodichloroethane	0.18	0.06	no
BHC-alpha	0.16	0.39	yes
Lindane	0.21	0.34	yes
BHC-beta	0.50	0.60	yes
Heptachlor	0.21	0.21	no
Aldrin	0.26	0.24	no
Heptachlor epoxide	0.26	0.40	yes
Chlordane	0.40	0.56	yes
o,p'-DDE	0.19	0.44	yes
Dieldrin	0.17	0.39	yes
Endrin	0.11	0.14	no
o,p'-DDT	0.09	0.18	yes
p,p'-DDD	0.10	0.27	yes
148 *p,p'*-DDT	0.08	0.17	yes

Table Contd.

	Probability Detectable in Groundwater	Probability Detectable in Surface Water	Significantly Different at 0.05 Level
Arsenic	1	1	no
Beryllium	1	1	no
Cadmium	1	1	no
Copper	1	1	no
Chromium	1	1	no
Nickel	1	1	no
Lead	1	1	no
Selenium	1	1	no
Zinc	1	1	no

Table 9.10. Organics in Drinking-Water Wells

Chemical	Highest Concentration (ppb)	State
Trichloroethane	27,300	PA
Toluene	6,400	NJ
1,1,1-Trichloroethane	5,440	ME
Acetone	3,000	NJ
Methylene chloride	3,000	NJ
Dioxane	2,100	MA
Ethyl benzene	2,000	NJ
Tetrachloroethylene	1,500	NJ

Synthetic organic chemicals, including most chlorinated hydrocarbons, can cause health problems ranging from acute effects from high doses, such as nausea, dizziness, tremors, and blindness to skin eruptions and central nervous system impairment at lower concentrations. CEQ suggested that ingestion of low concentrations of synthetic organic chemicals over an extended period of years may prove fatal. Studies of occupational 149 medicine, cancer studies using laboratory animals, and epidemiological studies were reviewed by CEQ to determine the possible long-term effects of such exposure.

Studies of occupational exposure have shown that some chemicals have adverse effects on human reproductive, central nervous, and other systems, but there is no direct human evidence to indicate whether exposure to such chemicals, at the highest concentrations found in groundwater, would represent a significant threat to human reproduction. DBCP, vinyl chloride, EDB, benzene, toluene, xylene, selected chlorinated ethanes and phthalate esters, PCBs, and the chlorinated dibenzo-p-dioxins are a few substances suspected of damaging the reproductive system or causing birth defects in humans.

Known or suspected carcinogens found in groundwater have been summarized and are presented in Table 9.11 . Thirty-two of the listed substances are organic chemicals and five are heavy metals. The Office of Technology Assessment (OTA) claims there is very little doubt that benzene, benzidine, inorganic arsenic, vinyl chloride, chromium, and nickel are human carcinogens. However, the latter two are not likely to be present in groundwater in their carcinogenic forms (hexavalent chromium, nickel and nickel salts) or at carcinogenic levels. Evidence from animal studies has also been reviewed, and the following substances are confirmed animal carcinogens: chlorinated aliphatic hydrocarbons (carbon tetrachloride, chloroform, TCE, PCE, and others) and chlorinated hydrocarbon pesticides (aldrin, chlordane, DDT, dieldrin, heptachlor, toxaphene, and others).

Many epidemiological studies have been conducted on the health effects resulting from ingestion of chlorinated water. The effects of chlorination are likely to be limited to their use in surface waters, however, because groundwater generally has a low humicacid content, and it is these acids that react with chlorine during the chlorination process to produce trihalogenated methanes.

Table 9.11. Known or Suspected Carcinogens Identified in Groundwater

Only substances classified as carcinogens by the International Agency for Research on Cancer, the Department of Health and Human Services, and the Environmental Protection Agency are listed. In all cases, the evidence has been subjected to peer review
I. Organic Chemicals Other Than Pesticides
Aromatic Hydrocarbons
Benzene

Table Contd.

Polynuclear Aromatic Hydrocarbons
Anthracene
Phenanthracene
Ethers
Dioxane
Halogenated Aliphatic Hydrocarbons
Carbon tetrachloride
Chloroform
1,2-Dichloroethane
1,1-Dichloroethane
Hexachloroethane
1,1,2,2-Tetrachloroethane
Tetrachloroethylene
1,1,1-Trichloroethane
1,1,2-Trichloroethane
Trichloroethylene
Vinyl chloride
Halogenated Aromatic Hydrocarbons
Polychlorinated biphenyls (PCBs)
Aromatic Amines
Aniline
Benzidine
Chlorophenols
1,4,6-Trichlorophenol
Esters
bis-(2-ethylhexyl)phthalate
Multifunctional or Miscellaneous Structural Types
Dichlorobenzidine
4,4-Methylene-bis-2-chloroaniline (MOCA)
2,3,7,8-Tetrachlorodibenzo-p-dioxin

II. Pesticides

Chlorinated Hydrocarbons
Aldrin
Chlordane
DDD
DDT
1,2-Dibromo-3-chloropropane (DBCP)
Dieldrin

Heptachlor
Kepone
Toxaphene

III. Inorganic Chemicals

Arsenic
Beryllium
Cadmium
Chromium
Nickel

Furthermore, groundwater from domestic wells, as opposed to municipal wells, usually does not undergo chlorination prior to use. The ecological epidemiological studies of surface-water ingestion suggest an association between either chlorination or the presence of trihalogenated methanes and cancer mortality in several areas of the United States. In five case studies in New York, Illinois, Wisconsin, Louisiana, and North Carolina the risk of rectal cancer was found to be higher for populations supplied by chlorinated water than for those supplied by unchlorinated water. Colon cancer risks were statistically significantly raised in three of the studies, and bladder cancer risks in two (New York and North Carolina). No clear trend was evident to determine whether cancer risks increased with increasing exposure to organic contaminants. The associations of certain types of cancer with the ingestion of chlorinated water are thought by many investigators to be non-specific and probably associated with something else beside water exposures.

A new area of study in which very little is yet known is that of the synergistic effects of contaminated groundwater. Synergistic effects in groundwater would involve interactions between chemicals and biological processes. OTA considers synergism as potentially one of the most important health issues concerning groundwater thus far. In addition, very little is known about the metabolism of many chemicals. It is known that some chemicals, such as dichloroethylene and dichloroethane, pass through reactive 152 intermediate stages in which they have a potential for damaging living tissue. The actual damage depends on the particular chemical, the rate of metabolism, and the ability of the organism to tolerate the presence of reactive intermediates.

Health Effects Due to Radionuclides in Groundwater

The effects on public health of high levels of radionuclides have been studied thoroughly for some time. Effects from low levels, however, are often more difficult to identify. Trace amounts of radioactive substances and their degradation products occur naturally in most groundwater; concentrations of these substances vary from area to area depending on the materials through which the groundwater has flowed. Man-made sources of radionuclides also contribute to the potential of groundwater contamination. Such sources include nuclear waste disposal sites, nuclear fallout, nuclear testing, nuclear accidents, waste tailings and piles, and mine drainage. Radium-226 is the most significant radionuclide found in groundwater; its presence is most often due to natural geologic conditions.

According to the National Academy of Sciences the risk of cancer from total average background radiation is estimated to be 4.5-45 fatalities per million persons per year (based on certain assumptions). Less than 1% of this total can be attributed to drinking contaminated water. Under average conditions, no measurable developmental or teratogenetic effects due to radiation are expected to occur from drinking groundwater containing radionuclides. However, in some areas in the Midwest -particularly Iowa, Illinois, Wisconsin and Missouri approximately 1 million people have been exposed to significant levels of radiation through groundwater. In Illinois and Iowa alone, approximately 500,000 people have water-well supplies containing radium-226 levels of 3-6 pCi/I, some 300,000 with levels of 6-9 pCi/I, and some 120,000 with levels of 9-80 pCi/I. Continuous consumption of water containing radium-226 or radium228 at 5 pCi/I may cause between 0.3 and 3 incidents of cancers per year per million exposed persons.

The lack of comprehensive nationwide surveys of the extent and severity of groundwater contamination and the paucity of groundwater contaminants that have actually been tested for carcinogenicity make it impossible to assess the national risk of 153 drinking groundwater. We do know, however, that in their pure form, at least some of the chemical contaminants are carcinogenic, that at high concentrations and in pure form some cause reproductive problems or affect the central nervous system or the cardiovascular system, and that in high concentrations some compounds cause poisoning and other problems such as skin rashes and diarrhea. We also know that pathogens probably cause more outbreaks of disease than are actually reported. Thus, adverse effects on human health have resulted from exposure to contaminated groundwater, but more research is required on the actual incidence and causal agents of disease, and a more rigorous inventorying of outbreaks of water-related disease would be very useful. Long-term monitoring of well water and the health of the population dependent on it might elucidate the cancer, cardiovascular, and other health risks and should be undertaken in the near future.

❑❑❑

10

Monitoring and Checking the Quality of Groundwater

Introduction

The costs of various methods, data management, and monitoring practices for disposal wells and concludes with illustrative examples. An earlier review, also put out by the General Electric Company, gives site-specific illustrative examples of monitoring practices for incidents involving oil-field brine disposal, plating-waste contamination, landfill-leachate contamination, oxidation ponds, and agricultural return-flow. The firm of Geraghty and Miller runs periodic seminars on the fundamentals of groundwater quality protection, in which monitoring practices are dealt with in detail. The Office of Technology Assessment report covers federal and state monitoring programs and reviews the general approach in conducting a hydrologic investigation. Gillham prepared a report for the American Petroleum Institute that covers groundwater monitoring procedures and points out sampling biases that may occur during monitoring.

EPA Ground-Water Monitoring Strategy

The Office of Drinking Water, Office of Solid Waste, Office of Emergency and Remedial Response, Hazardous Waste Ground-

Water Task Force, Office of Pesticide Programs, and Office of Management Systems and Evaluation prepared the strategy cooperatively under the leadership of the Office of Ground Water Protection. Together they analyzed and summarized the need for and use of monitoring data. They determined that in order to develop useful groundwater protection programs, information on a broad range of environmental problems must be acquired. Information must be gathered by federal, state, and local officials on the causes of these problems, their public health implications, and their potential for control.

The monitoring strategy focuses on such fundamental activities of groundwater monitoring as improving the quality and management of groundwater data collected and analyzed by or on behalf of EPA programs, and upgrading existing agency monitoring activities and filling the gaps between them.

The strategy is designed to: 1) be broad in scope and not address specific details of any specific monitoring program; 2) focus on a five-year horizon for implementation, with both short- term and long-term action items; 3) recognize that the job is complex and that EPA and others are interested in considering what constitutes an appropriate overall groundwater monitoring program; 4) provide a balance in the strategy-implementation plan between general action items and a detailed work plan; and 5) focus on solutions to the groundwater monitoring problems identified in the Office of Technology Assessment (OTA) report *Protecting the Nation's Ground Water from Contamination*, and not merely restate their analyses.The Ground-Water Monitoring Strategy recognizes and responds to a number of needs identified in the 1984 EPA Ground-Water Protection Strategy. These are:

1. The need to strengthen state programs by assisting state decision makers in developing and gaining access to needed groundwater monitoring data.
2. The need to deal with contamination sources of national 231 concern. The strategy addresses ways to acquire information on the extent and consequences of the many sources of groundwater contamination of national concern, so that decision makers may protect the public and the environment.
3. The need to create a policy framework. In order to implement consistency in a groundwater protection policy centered on

differential protection, site specific and regional groundwater data are needed.

4. The need to strengthen EPA's internal groundwater organization. The strategy will help EPA by reviewing (a) its approach to monitoring, (b) the need for and use of groundwater monitoring data, and (c) how that data can be made accessible to other environmental decision makers.

The monitoring strategy addresses seven monitoring objectives:

1. To characterize the nation's groundwater resources which entails collecting background data on groundwater resources and their sellings.
2. To identify new contamination problems which will provide information on previously unidentified threats to groundwater quality.
3. To assess known problems to support regulatory development and standard setting and respond to site-specific problems - which involves monitoring sites with previously identified contamination in order to support regulatory and standard - setting requirements and to respond to site-specific problems.
4. To ensure compliance with regulations which entails the collection of monitoring data so that program managers, permit writers, inspectors, and enforcement officials responsible for controlling or abating the consequences of contamination can ensure compliance.
5. To evaluate program effectiveness which entails the collection of groundwater monitoring data to evaluate regulatory effectiveness in protecting groundwater quality.
6. To improve data quality which involves improving quality assurance and quality control procedures.
7. To develop a groundwater data management system — which includes activities such as determining what data is to be collected, developing protocols on data entry, and making the data accessible to users.

Federal Statutes Requiring Groundwater Monitoring

There are 10 Federal regulations that require monitoring of groundwater:

1. Atomic Energy Act – requires low level radioactive facilities to conduct pre-operational, operational, and post-operational monitoring related to the development of geologic repositories; and as part of remedial action programs at storage and disposal facilities for radioactive substances.
2. Clean Water Act – requires treatment facilities to monitor on a case-by-case basis sites of land application of wastewater and sludge from sewage treatment plants.
3. Comprehensive Environmental Response, Compensation and Liability Act monitoring may be conducted by a state or by EPA in response to the release of any hazardous substance, contaminant, or pollutant (as defined by CERCLA).
4. Federal Insecticide, Fungicide and Rodenticide Act monitoring may be conducted by EPA in instances when certain pesticides are contaminating groundwater.
5. Federal Land Policy and Management Act– requires facilities to monitor geothermal recovery operations on federal lands for at least one year prior to production.
6. Resource Conservation and Recovery Act – requires facilities to monitor all hazardous waste land disposal facilities. Background quality, detection, compliance, and corrective-action monitoring are all required.
7. Safe Drinking Water Act – monitoring may be required for injection wells used for in-situ or solution mining of minerals (Class III wells) where injection is into a formation containing less than 10,000 mg/l TDS; monitoring may be required for wells injecting beneath the deepest underground source of drinking water (Class I).
8. Surface Mining Control and Reclamation Act – requires facilities to monitor surface and underground coal-mining operations to determine the impacts on the hydrologic balance of the mining and adjacent areas.
9. Toxic Substances Control Act – requires facilities to monitor prior to the disposal of PCBs.
10. Uranium Mill Tailings Radiation Control Act – requires facilities to monitor active mill tailing sites. Background quality, detection, compliance, and corrective-action monitoring are all required; monitoring may be required for inactive sites if contamination has occurred.

Interstate compactssuch as those for the Delaware River Basin and the Susquehanna River Basin have the authority to monitor groundwater quality and to formulate and carry out plans to protect groundwater quality.

Federal, State, and Private Organizations that Carry Out Monitoring Activities

Federal, state, and private organizations use several different monitoring approaches to determine groundwater contamination. These include:

- Ambient-trend monitoring measures quality in relation to standards and reflects temporal and spatial trends in a groundwater area.
- Source monitoring measures effluent quality and quantity from pollution sources that could affect groundwater. Present methodology is usually geared toward this type of monitoring and concentrates on the most important sources.
- Case-preparation monitoring is used to accumulate data for enforcement actions.
- Research monitoring is used for studies of groundwater quality and pollution occurrence and movement.

Federal

The Water Research Division of the USGS is currently the principal federal division conducting groundwater research. The largest program, the Federal-State Cooperative Program, conducts groundwater investigations that involve the collection of hydrologic data and the investigation of water resources relevant to state and local needs. Within this program, costs are shared equally by the USGS and state and local agencies. This program is active in all 50 states.

The USGS has also proposed the National Water Quality Assessment Program (NAWQAP), which will attempt to describe water quality on a national scale. Other USGS programs involving groundwater monitoring include the Regional Aquifer System Analysis Program (RASA); the National Water Resources Conditions Report, which is published monthly, the Toxic Waste Ground-Water

Contamination Program; the Radioactive Waste Program; and the Coal Hydrology and Oil Shale Hydrology Programs.

The EPA requires resource assessment information to support most of its activities. For example, monitoring is conducted on groundwater at Superfund sites and RCRA sites. The EPA also is involved with the identification and assessment of new groundwater problems. RCRA and Superfund programs conduct investigations on new or additional problems concerning groundwater contamination. Studies conducted under the Clean Water Act and the Safe Drinking Water Act have also been used to identify undiscovered contamination problems. Recently, the National Pesticide Monitoring Plan has been working on the identification of problems associated with pesticide use and exposure-related illnesses. To address known goundwater problems, several surveys and studies are now being conducted by the Office of Pesticides Programs (OPP) and the Office of Drinking Water (ODW) on pesticide contamination, by the ODW on inorganics and radionuclides, by the OTS on leaking underground storage tanks, by RCRA on non-hazardous landfills, and by the Office of Water on publicly owned treatment works/lagoons. Other federal agencies involved with groundwater monitoring are the Bureau of Reclamation, the Army Corps of Engineers, the Soil Conservation Service, the Nuclear Regulatory Commission, the Office of Surface Mining, the Department of Defense, the Department of Agriculture, and the Department of Energy.

Basic Procedures in Setting Up a Groundwater Monitoring Program

As factors such as climate, population, land use, geohydrology, and pollution sources vary from site to site, the design of an appropriate monitoring program should vary accordingly. A single set of guidelines for all monitoring programs would not suffice unless it were very general and flexible. Each program should reflect those specific variables identified at each site. While designing monitoring systems, five general questions should be kept in mind:

- What information is required?
- What techniques are applicable?
- What should be the numbers and locations of measuring points?

- How frequently should samples be collected?
- How much money can be spent?

During all monitoring procedures, steps must be taken to limit any actions that may alter the physical and chemical environment of the samples. These changes may occur from:

- physical disturbance during drilling, such as vertical mixing between contaminated and uncontaminated zones
- contamination caused by foreign materials used during drilling operations
- contamination by foreign materials such as dust or soil in the sampling installation or equipment
- contamination caused by materials of the sampling installation, sampling equipment, and sample containers
- adsorption onto materials in contact with the sample
- cross contamination between sampling points
- chemical alterations caused by degassing and volatilization
- direct or indirect chemical changes occurring as a result of atmospheric contamination of the sample

The following is a very brief and general outline of the steps that should be taken in implementing a groundwater monitoring program.

Initial Site Assessment

Vital hydrologic and groundwater quality information can be obtained through visual investigation, research of hydrologic and land use information, and in-situ tests of the site in question. Simply through visual examination of the site, hydrologic as well as contaminant information can be assessed. The location of hills and streams can help in determining hydrologic flow, just as areas of distressed vegetation, soil stains, and refuse can help in estimating groundwater contamination movement. Hydrologic and land use reports, topographic maps, and aerial photographs can also be of major importance in determining groundwater flow and potential sources of pollution. More popular methods that provide for a more detailed, site-specific data set are geophysical testing, temperature studies, tracer studies, soil boring and well drilling programs, and trench and excavation work. Geophysical testing methods include:

- electrical earth resistivity and electomagnetic terrain conductivity techniques to evaluate depth to water and areal extent of contaminants involving dissolved ions;
- borehole logging to determine vertical limits of aquifers and ionized contaminants;
- seismic surveys to identify subsurface geological conditions;
- proton magnetometers to locate buried metallic objects; and
- Soil vapor (soil sniffing) surveys to identify areas of contamination with volatile and semi-volatile organic compounds.

Selecting Parameters

All known or suspected contaminants should be identified and considered for analysis. The constituents in industrial and municipal wastewater that may affect groundwater. The substances chosen should be limited to those for which approved analytical methods exist. In the absence of known or suspected contaminants, substances from the area's mineralogy, substances listed in the drinking water standards list, or EPA's priority pollutants may be tested. Where cost is a consideration, which is in most cases, testing should be limited to the most mobile contaminants.

Types of Wells

Groundwater sampling wells can be divided into two general types: single level installations, which are the most common type, and multilevel installations. Single level installations consist of one single inlet interval for each borehole. This type of well provides no information on the vertical distribution of contaminants. Multilevel installations are used for more detailed mapping of contaminated areas. They consist of either a bundle of small diameter tubes buried together in a single borehole, or a single well with a long screened interval designed for the collection of samples at various depths.

Table 10.1. Constituents in Industrial and Municipal Wastewater Having Significant Potential for Groundwater Contamination

Mining		
Metal and Coal Mining Industry		
pH	Zinc	Magnesium
Sulfate	Tin	Silver
Nitrate	Vanadium	Manganese
Chloride	Radium	Calcium
Total dissolved solids	Phenol	Potassium
Phosphate	Selenium	Sodium
Copper	Iron	Aluminum
Nickel	Chromium	Gold
Lead	Cadmium	Fluoride
Uranium	Cyanide	
Paper and Allied Products		
Pulp and Paper Industry		
COD/BOD	Phenols	Nitrogen
TOC	Sulfite	Phosphorus
pH	Color	Total dissolved
solids		
Ammonia	Heavy metals	Biocides
Chemicals and Allied Products		
Organic Chemicals Industry		
COD/BOD	Alkalinity	Phenols
pH	TOC	Cyanide
Total dissolved solids	Total phosphorus	Total nitrogen
	Heavy metals	
Inorganic Chemicals, Alkalies, and Chlorine Industry		
Acidity/alkalinity	Chlorinated	Chromium
Total dissolved solids	benzenoids and	Lead
Chloride	polynuclear	Titanium
Sulfate	aromatics	Iron
COD/BOD	Phenols	Aluminum
TOC	Fluoride	Boron
	Total phosphorus	Arsenic
	Cyanide	
	Mercury	
Plastic Materials and Synthetics Industry		
COD/BOD	Phosphorus	Ammonia
pH	Nitrate	Cyanide

Table Contd.

Phenols	Organic nitrogen	Zinc
Total dissolved solids	Chlorinated benzenoids and polynuclear aromatics	Mercaptans
Sulfate		
Nitrogen Fertilizer industry		
Ammonia	Sulfate	COD
Chloride	Organic nitrogen compounds	Iron, total
Chromium		pH
Total dissolved solids	Zinc	Phosphate
Nitrate	Calcium	Sodium
Phosphate Fertilizer Industry		
Calcium	Acidity	Mercury
Dissolved solids	Aluminum	Nitrogen
Fluoride	Arsenic	Sulfate
pH	Iron	Uranium
Phosphorus	Cadmium	Vanadium
	Radium	
Petroleum and Coal Products		
Petroleum Refining Industry		
Ammonia	Chloride	Nitrogen
Chromium	Color	Odor
COD/BOD	Copper	Total phosphorus
pH	Cyanide	Sulfate
Phenols	Iron	TOC
Sulfide	Lead	Turbidity
Total dissolved solids	Mercaptans	Zinc
Primary Metals		
Steel Industry		
pH	Cyanide	Tin
Chloride	Phenols	Chromium
Sulfate	Iron	Zinc
Ammonia	Nickel	
Electric, Gas, and Sanitary Services		
Power Generation Industry		
COD/BOD	Copper	Phosphorus
pH	Iron	Free chlorine
Polychlorinated biphenols	Zinc	Organic biocides
	Chromium	Sulfur dioxide
Total dissolved solids	Other corrosion inhibitors	Heat
Oil and grease		

Municipal Sewage Treatment		
pH	Nitrate	Sulfate
COD/BOD	Ammonia	Copper
TOC	Phosphate	Lead
Alkalinity	Chloride	Tin
Detergents	Sodium	Zinc
Total dissolved solids	Potassium	Various organics

Table 10.2 Cost of Chemical Determinations of Water Samples

Parameter	Method of Analysis	Price per Sample ($)
Metals and Minerals		
Aluminum	Atomic Absorption – Flame	13
Antimony	" " Furnace	13
Arsenic	" Furnace	13
Barium	" Flame	13
Beryllium	" " Furnace	13
Boron	" " Flame	13
Cadmium	" " Furnace	13
Calcium	" " Flame	13
Chromium	" " Furnace	20
Cobalt	" " Furnace	13
Copper	" " Furnace	13
Gold	" " Flame	13
Iron	" " Flame	13
Lead	" " Furnace	13
Magnesium	" " Flame	13
Manganese	" " Flame	13
Mercury	" " Cold vapor	20
Molybdenum	" " Furnace	13
Nickel	" " Furnace	13
Palladium	" " Flame	13
Phosphorus	" " Furnace	13
Platinum	" " Flame	13
Potassium	" " Flame	13
Selenium	" " Furnace	13
Silicon	" " Flame	13
Silver	" " Furnace	13
Sodium	" " Flame	13
Thallium	" " Furnace	13
Tin	I" Furnace	13
Titanium	" " Furnace	13
Uranium	" " Flame	20

Table Contd.

Parameter	Method of Analysis	Price per Sample ($)
Vanadium	" " Furnace	13
Zinc	" " Flame	13
Drinking Water and Waste Water		
Acidity	Electrometric titration	12
Alkalinity – total	Auto analyser colorimetric (methyl orange)	15
– phenolphthalein	Titration	12
– bicarbonate	"	12
– carbonate	"	12
– hydroxide	"	12
Bromide	Solvent extraction with colorimetric development	25
Carbo dioxide – free	Titrimetric	12
– total	"	12
Carbon – total	Combustion	25
– inorgainc	"	25
– organic	"	25
Chemical oxygen demand	Low valve COD ampule method – water	25
	– sediment	35
Chloride	Auto analyzer colorimetric ferricyanide	15
Chlorine – total residual	Amperometric titrimetric	10
– free available	" "	10
Chlorophyll A,B,C, Phaeopigments	Spectrophotometer, acetone extraction	40
Hexavalent chromium	APDC/MIBK extraction with graphite	20
	furnace atomic adsorption quantitation	
Color	Visual comparison platinum cobalt	9
Specific conductance	Conductance cell, Wheatstone bridge	8
Cyanide	Auto analyzer distillation water	25
	sediment	35
Fluoride	Potentiometric, ion selective electrode	12
Specific gravity	Method #210	12
Total hardness	Auto analyzer (as CaCO3)	15
Iodide	Titrimetric	12
MBAs (surfactants)	Extraction and colorimetric spectrophotometer	30
Odor	Threshold measurement	12

Parameter	Method of Analysis	Price per Sample ($)
Oil and grease	Gravimetric separatory	25
	funnel extraction water sediment	35
Dissolved oxygen	Electrode	Field time charge
Particle size distribution	Sieve analysis test	Variable
pH	Electrometric	5
Phenolics	Auto analyzer distillation water	25
	Sediment	35
Total residue	Gravimetric	12
Dissolved residue	Gravimetric	12
Total non-filterable residue	Gravimetric	12
Volatile residue	Gravimetric	12
Reactive silica	Colorimetric	12
Sulfate	Auto analyzer barium chloride	12
Sulfide	Titrimetric (iodine)	12
Temperature °C	Calibrated glass thermometer	Field time charge
Turbidity	Nephelometric	12
Nutrient Parameters		
Ammonia nitrogen	Auto analyzer colorimetric phenate	15
Total Kiehldahl nitrogen	" " " "	25
Nitrate nitrogen	" " " "	15
Nitrite nitrogen	" " " "	15
Organic nitrogen	" " " "	30
Ortho phosphorus	Auto analyzer colorimetric molybdate	15
Dissolved ortho phosphorus	Color development w/ spectrophotometer	15
Total soluble phosphorus	" " " "	12
Total phosphorus	Auto analyzer colorimetric molybdate	12
Sanitary Parameters		
Fecal coliform bacteria	Membrane filter	15
Total coliform bacteria	Membrane filter	15
Fecal streptococci	Membrane filter	15
Biochemical oxygen demand	5-day incubation	30
Organics		
Trihalomethanes	Gas chromatography/mass s pectrometry	150
Chloroform	(GC/MS) – water	
Dibromochloromethane		
Bromodichloromethane		

Table Contd.

Parameter	Method of Analysis	Price per Sample ($)
Bromoform		
Polychlorinated biphenyls (PCBs)	Gas chromatography water	80
	sediments	85
Polybrominated biphenyls (PBBs)	Gas chromatography water	75
	Sediments	85
Herbicides	Gas chromatoraphy water	185
(2,4-D/2,4,5-TP, Silvex)	Sediments	220

Table 10.3. Cost of Priority Pollutants Analysis (in $)

Samples per Batch	Volatile	Acid Extractables Base/Neutral Extractables		Pesticides and PCBs	Total
1-10	235	155	225	155	770
10	215	140	205	140	700

Samples per Batch	13 Metals and Total Cyanides	PCE/TCE
1-10	120	48
10	110	42

Monitoring Network

The validity of a monitoring effort depends upon the extent to which point measurements are representative of the spacial distribution of contamination. The location and number of wells must be compatible with the objectives of the program, the physical characteristics of the hydrogeologic environment, and the physical and chemical characteristics of the contaminant species. At most monitoring sites a single monitoring well cannot adequately determine the extent of contamination; it is necessary to have several wells at different depths. Groups or clusters of wells, each in separate boreholes, are referred to as piezometer nests. Several nests at a site can provide a three-dimensional representation of groundwater flow and chemical composition of the groundwater. Werth discourages the use of linear alignment of wells unless, for example, the flow along identified bedrock fractures is to be investigated. It is also suggested that one well be installed off site to provide background water-quality information for the area examined.

Well Drilling Techniques

Drilling methods should reflect the well construction details, the soil sampling requirements, and the proposed water-quality sampling program. Werth and Gillham have listed several well drilling methods and point out the positive and negative aspects of each:

1. Shallow monitor wells are usually constructed with an *auger*. This technique is simple and cheap, requires no drilling fluids, and provides good soil samples from core materials. However, both the well depth and diameter are limited, contamination is possible from the metal components of the drill, and the method can only be used in unconsolidated geologic material.
2. When using a *mud rotary*, drillers are not limited to any specific depth or geologic material, and undisturbed core materials can be obtained for sampling. However, the method is slow going, contamination from the metal parts is possible, and fluids are used in the process.
3. *Air rotary* rigs are quite fast. This method is not limited to depth or geologic material, and drilling fluids are not necessary. However, the method is expensive, the air used in the drilling process may contaminate the samples, and it is not a suitable method in areas contaminated by toxics.
4. *A cable-tool* rig can be used to hammer down a well shaft. It provides good geologic samples, requires little or no drilling fluid, and is not limited to certain geologic materials. However, it is slow.
5. The *barber* rig is a very good method for drilling wells. It is 250 rapid, provides good soil and water samples, and uses very little or no fluids in the process. However, in 1985, there were only three of these rigs in the United States.
6. *Diamond drilling* is a good technique for drilling into crystalline rock. It provides good geologic samples, and there is no depth limit. However, it is used only in consolidated rock or compacted tills; foreign water is used during the process, which may contaminate the formation.

All drilling processes have the potential to physically and chemically contaminate samples through both disturbing the hydrogeologic conditions of the subsurface, and through the use of drilling equipment and fluids. There are no general guidelines for the prevention of contamination because of the wide variety of methods and tools used. However, preventive measures can be taken according to the specifics of each case.

Casing Materials

The types of well casings and screens used for monitoring depend upon which contaminants are being monitored and how the wells are to be set in place. Most important, the materials chosen should demonstrate resistance to corrosion from the suspected contaminants. Werth lists six materials from which the screens and casings are made and the pros and cons of using each:

1. PVC — is inexpensive and easy to work with and reasonably durable; however, it is not totally inert.
2. Steel — is also inexpensive and strong; however, it sorbs organics and corrodes readily.
3. Galvanized steel — is inexpensive, reasonably resistant, and strong, but it leaches some metals.
4. Stainless steel — is very strong and resists corrosion; however, it is expensive to use and leaches some metals.
5. Eflon — is inert and corrosion proof, but it is expensive and not a very strong material.
6. Fiberglass — is corrosion resistant, but is not very strong.

Glues and solvents, applied to join PVC pipes together or cloth filters to well casings, should not be used with monitoring wells, for they may contaminate samples.

Annular seals should be placed around all monitoring wells. They are used to guard against infiltration of surface waters to the well screens and to protect the integrity of the confining unit 251 penetrated by the well.

Sampling Procedures

There are a large number of sampling devices available for monitoring groundwater. Their principal operating methods include: down-hole collection (or boring), suction lift, positive displacement, gas-drive, gas-lift, jet-pumps, and destructive sampling. Careful measures should be taken throughout sampling procedures to prevent bias in sampling results. Potential bias can occur from sample contact with foreign materials used with the sampling device, cross contamination between installations, chemical alterations as a result of degassing or atmospheric contamination, and sampling stagnant water. Pressure changes affect CO_2 and pH concentrations. Oxygen and temperature changes alter the solubilities of solids and gases, and also alter biological activity. These changes can be minimized if samples are pumped directly into units containing monitoring devices.

Evaluation of Data

Reviewing and interpreting monitoring results may enable quality trends, new problems, improvements, and effectiveness of remedial actions to be determined. Methods of groundwater and contaminant modeling may be implemented to interpret these results. Assessment of substandard water occurrences and prediction of future alteration in quality due to population and land use changes should also be considered. The final step in a well-organized monitoring program is the dissemination of the results in a readily understandable and usable form.

Variabilities involved in the groundwater monitoring process must be taken into account when evaluating groundwater data, Saar lists five factors in monitoring that can lead to inaccurate data results: actual variability in the environment; variability resulting from monitoring system setups; variability introduced during sampling; variability caused by 252 laboratory procedures; and lastly, variability introduced during calculations and reporting of data.

Monitoring Costs

Depending on several interrelated factors, the costs of a monitoring program may vary considerably from case to case. The two key factors are the type and quality of survey used to locate and identify the contamination and the type and number of monitoring wells. These, in turn, are influenced by site size, geophysical aspects such as depth and type of formations, and often by whether a public agency or the site owner is requesting the program. Surveys of the contaminated site may vary from a preliminary assessment describing whether contamination is present and the general direction of groundwater flow, to a detailed assessment describing the distribution and quantity of each contaminant and the quantity, rate, and direction of the groundwater flow. The monitoring wells vary in configuration and construction, and both are described earlier in this chapter. It has been estimated that 1% of the total capital costs would provide adequate maintenance for the wells. Where a typical monitoring well for a waste-disposal site may cost several thousand dollars, wells monitoring the underground injection of contaminants may cost ten times that much. Monitoring programs for underground injection wells will vary considerably according to formation type, depth, and the location of any hydrologically connected aquifers that also must be monitored. For any monitoring program, costs can only be judged on a case-by-case basis.

Data Management and Information Retrieval

There are several computerized information storage systems suitable for groundwater. The STORET system, developed by the U.S. Public Health Service and now operated by EPA, has a waterquality file and a general point-source file. It was developed to help ensure compliance with the Federal Water Pollution Control Act Amendments. This data base contains water-quality data for both surface waters and groundwaters. It includes chemical, physical, biological, and radiochemical water analyses, as well as many other types of information. STORET receives data from several federal agencies, from more than 40 state agencies, from USGS-state cooperative investigations, and, on a monthly basis, from USGS WATSTORE. EPA's Office of Pesticides is currently investigating the use of STORET for the National Pesticides Monitoring Program. RCRA and the Superfund program are also investigating the use of STORET.

Drawbacks to the STORET system result from a lack of quality control over the data entered and an absence of hydrologic and well information associated with water quality data. In other words, much of the data received reflects the needs of the investigation from which the data has come, which results in a wide variety of information.

WATSTORE, the National Water Data Storage and Retrieval System, established in 1971, is operated by the USGS. Groundwater data is found in the Ground-Water Site Inventory file (GWSI) of the WATSTORE system. GWSI contains an inventory of wells, springs, and other sources of groundwater, and includes data on site locations, geohydrologic characteristics, well construction history, and various field measurements. This data is collected from USGSstate cooperative investigations, and other USGS, state and local groundwater investigations.

WATSTORE information is limited in that the data is derived, for the most part, from investigations of point sources, and most of the information is related to inorganic contamination.

NAWDEX, The National Water Data Exchange, operated by the Water Resources Division of the USGS, is an index of water data locations used by the USGS and is designed to assist users of water data with the identification, location, and aquisition of needed data. The USGS Water Resources Scientific Information Center (WRSIC) has computerized abstracts and bibliographic information on water related scientific literature and maintains a data base on water research in progress. In addition, the USGS is now developing a National Water Information System (NWIS), incorporating WATSTORE, WRSIC, NAWDEX, and other computerized bibliographic information into a master file.

❑❑❑

11

WATER POLLUTION FROM PULP AND PAPER COMPANIES

Introduction

Pollution performance of the pulp and paper companies, based on an aggregation of the mill pollution performance, is presented. Information is presented first for each of the three pollutants discussed and then for the overall pollution performance of the firm.

Measurement and Rankings

Pulp and paper mills operate in different parts of the United States, using different processes in making pulp and paper, and the wastewater empties into rivers, streams and lakes of different sizes. Thus, it is difficult to compare these mills either individually or in aggregate. Nevertheless, all the managers of these mills should have been aware of the goals of the Federal Water Pollution Control Amendments to make all lakes, streams, and rivers swimmable or fishable and to achieve no pollution discharge. Furthermore, the technology to clean up the mill to a near-zero level of pollution was available to the firms. Therefore, the level of pollution that the mill emitted was a decision that was controlled by the firm's managers.

In order to compare companies that have mills of different sizes it was necessary to calculate the pollution emissions per ton of production. The overall water pollution performance of each of the companies can also be determined. However, this requires that the relative detriments of each pollutant be known. That is, is pollution that results in a high BOD measure more damaging than either TSS emissions or emissions that cause improper pH? Since each of these elements of water pollution is harmful, it was decided to weight them equally.

The company with the highest emissions for a given pollutant was given a score of 100 for that pollutant and all other measures of that pollutant were then scored relative to that company. This was done for each of the companies and for each measure of water pollution. Under the equal weighting scheme, the scores were then summed to arrive at an overall score. The firm with the highest score was the worst polluter. If another weighting scheme is utilized, the relative weights will have tc be factored in the process of combining the scores for each measure of pollution.

BOD Measures

The BOD measure for the total of the companies included in this study decreased dramatically over the 1998-2006 period. Table 11.1 summarizes the BOD measure for each company and for all the companies included in the analysis.

All the companies that had a worse-than-average measure of BOD in 1998 improved their performance by at least 43% over the 1988-96 period. The only companies that had an increase over the period (Hammermill, Mead and Westvaco) were much better than average in 1988. In 1996, however, International Paper had a BOD measure that was clearly the worst of all the companies in the study.

An examination of the worst mills for 1996 shows that there were only six mills that were worse than International Paper's measure of 12.11 pounds of BOD per ton of production. Had IP's mills daily production been known, several of its mills would be included in this table. What is striking about Table 11.2 is that the five worst mills are located in the state of Washington. Washington is highly dependent cn the forest industry for its economic well-being, and it appears to be allowing mills to pollute more than are other states.

The mills with the lowest BOD measures have all managed to control these pollution measures.

Table 11.1. Comparison of BOD Measures by Company

Company	(pounds/ton of production) 1988	1993	1996	% Change 1988-96
Boise Cascade	6.47	5.45	3.83	-41
Champion Int'l.	3.97	.89	.81	-80
Crown Zellerbach	21.40	16.31	8.13	-62
Georgia Pacific	16.23	10.83	6.92	-57
Great N. Nekoosa	10.88	4.20	1.83	-83
Hammerhill	2.38	6.72	5.91	+148
Int'l. Paper	21.32	7.81	12.11	-43
Kimberly Clark	7.01	5.71	-19	
Mead	1.28	3.80	2.30	+78
Potlatch	11.98	4.63	2.37	-80
St. Regis	2.56	1.52	1.16	-55
Scott	8.03	11.26	5.92	-26
Union Camp	4.24	3.61	-15	
Westvaco	1.51	1.62	1.92	+27
Weyerhaeuser	12.08	7.18	6.84	-43
Totals	10.47	5.74	4.45	-57

Table 11.3 shows that these nine mills all had a net discharge BOD measure of zero. Three of the mills are owned by Kimberly Clark and six of the nine mills are located east of the Mississippi River. However there appears to be no glaring reason why these mills have been completely successful in controlling the BOD measure while other mills do so poorly.

Table 11.4 scores and ranks the companies based on their BOD measure for 1978, 1983 and 1986. The information echoes Table 11.1. That is, overall the companies have greatly improved their performance between 1978 and 1986. Great Northern Nekoosa made relative improvements each year, going from 8th to 5th and then 3rd. Champion also had a major transition, from 5th in 1988 to 1st in 1993 and 1996.

Table 11.2. Mills with the Highest BOD/Ton of Production Measure, 1996

Mill	Location	Company	(pounds/ton of production) 1996
Cosmopolis	Washington	Weyerhaeuser	51.19
Bellingham	Washington	Georgia Pacific	26.41
Steilacoom	Washington	Boise Cascade	16.27
Camas	Washington	Crown Zellerbach	15.82
Longview	Washington	Weyerhaeuser	14.36
Lyons Falls	New York	Georgia Pacific	14.12

Table 11.3. Mills with the Lowest BOD Measures, 1996

Mill	Location	Company	(lbs/ton of production) 1996
Ashdown	Arkansas	Great N. Nekoosa	0
Beech Island	South Carolina	Kimberly Clark	0
Bucksport	Maine	Union Camp	0
Eastover	South Carolina	Union Camp	0
Hamilton	Ohio	Champion	0
Luke	Maryland	Westvaco	0
Memphis	Tennessee	Kimberly Clark	0
Neenah	Wisconsin	Kimberly Clark	0
St. Helens	Oregon	Boise Cascade	0

Table 11.4. Companies Ranked on the BOD Measure for 1988, 1993 and 1996

	1988		1993		1996	
Company	Score	Rank	Score	Rank	Score	Rank
Boise Cascade	30.23	6	25.49	8	17.91	8
Champion	18.54	5	4.16	1	3.79	1
Crown Zellerbach	100	13	76.23	15	38.01	14
Georgia Pacific	75.84	11	50.60	13	32.35	13
Great N. Nekoosa	50.85	8	19.64	5	8.57	3
Hammermill	11.12	3	31.41	9	27.60	10
Int'l. Paper	99.63	12	36.49	12	56.58	15
Kimberly Clark			32.74	10	26.72	9

	1988		1993		1996	
Company	**Score**	**Rank**	**Score**	**Rank**	**Score**	**Rank**
Mead	6.02	1	17.76	4	10.74	5
Potlatch	56.00	9	21.62	7	11.07	6
St. Regis	11.95	4	7.12	2	5.44	2
Scott	37.52	7	52.62	14	27.67	11
Union Camp			19.81	6	16.89	7
Westvaco	7.06	2	7.57	3	8.97	4
Weyerhaeuser	56.46	10	33.56	11	31.98	12

The companies that did relatively less well actually reduced their BOD measure, but not to as great a degree as the other companies. For example, Scott went from 7th to 14th and then to 11th place. However, its score went from 37.52 in 1978 to 27.67 in 1996. As indicated earlier, the one company that seemed to be doing considerably worse in controlling its BOD measure was International Paper and even that company improved its performance over the period examined.

Total Suspended Solids

The companies also made major strides in reducing TSS emissions over the 1988-96 period. As Table 11.5 indicates, most of the companies reduced their TSS emissions from 1988 to 1996.

Table 11.5. Comparison of TSS Emissions by Company

	(lbs/ton of production)			%Change
Company	**1988**	**1993**	**1996**	**1988-96**
Boise Cascade	10.25	7.59	4.42	-57
Champion Int'l	5.68	1.57	.41	-93
Crown Zellerbach	17.39	13.42	38.30	+120
Georgia Pacific	6.04	13.91	8.23	+36
Great N. Nekoosa	6.79	3.49	1.67	-75
Hammermill	.53	3.39	2.29	+333
Int'l. Paper	24.02	11.94	17.74	-26
Kimberly Clark	3.42	2.26	-34	
Mead	1.81	7.77	5.64	+212

Table Contd.

Company	(lbs/ton of production) 1988	1993	1996	%Change 1988-96
Potlatch	9.01	5.45	4.88	-46
St. Regis	3.40	.99	.81	-76
Scott	13.82	10.76	8.27	-40
Union Camp	2.09	3.11	+49	
Westvaco	.56	1.70	3.34	+493
Weyerhaeuser	10.39	5.21	6.86	-34
Totals	9.85	5.79	6.36	-35

However, for six companies, the performance deteriorated. Furthermore, from 1993 to 1996, the companies as a whole increased their TSS emissions. Of the six companies who had deteriorating performance in 1986 compared to 1988, only the performance of Crown Zellerbach is alarming. The second worst polluter in 1988, it proceeded to increase its pollution by 120% in 1996. Except for Georgia Pacific, the other companies that increased their pollution made almost no TSS net discharges in 1988.

An examination of the mills that emitted the most TSS per ton of production in 1996, as displayed in Table 11.6, reveals that Crown Zellerbach and Georgia Pacific own most of the worst mills. However, since mill production data was not available for International Paper (and Westvaco), and since

Table 11.6. Mills with the Highest Level of TSS Emissions, 1996

Mill	Location	Company	(lbs/ton of production) 1996
St. Francisville	Louisiana	Crown Zellerbach	105.1
Cosmopolis	Washington	Weyerhaeuser	61.4
Bellingham	Washington	Georgia Pacific	39.2
Port Angeles	Washington	Crown Zellerbach	29.7
Longview	Washington	Weyerhaeuser	20.6
Woodland	Maine	Georgia Pacific	18.6
Camas	Washington	Crown Zellerbach	17.3
Lyons Falls	New York	Georgia Pacific	17.3
Wauna	Oregon	Crown Zellerbach	15.0

International Paper had the second worst level of TSS emissions, it is probable that several of its mills should also be included. As was the case with BOD, the state of Washington is well represented on this list, another indication that Washington allows more water pollution in this industry than do other states.

What is disturbing about the performance of the worst mills in abating TSS pollution is that so many had achieved a level of no net TSS discharge in 1996. As Table 11.7 shows, 21 mills had no TSS net discharge in 1996. Twelve of the 15 companies are represented by these mills. What appears obvious is that Kimberly Clark dominates the list, and whatever technology it is using to abate TSS should be utilized by the other companies. Furthermore, the mills in Wisconsin have been successful in eliminating TSS discharge. This should be contrasted with the performance of the mills in Washington.

Table 11.7. Mills with the No Net TSS discharge, 1996

Mill	Location	Company	(lbs/ton of production) 1996
Ashdown	Arkansas	Great Nekoosa	0
Beech Island	South Carolina	Kimberly Clark	0
Brainerd	Michigan	Potlatch	0
Courtland	Alabama	Champion	0
Eastover	South Carolina	Union Camp	0
E. Millinocket	Maine	Great N. Nekoosa	0
Franklin	Virginia	Union Camp	0
Hamilton	Ohio	Champion	0
Kaukana	Wisconsin	Hammermill	0
Lakeview	Wisconsin	Kimberly Clark	0
Lee	Massachusetts	Kimberly Clark	0
Marinette	Wisconsin	Scott	0
Memphis	Tennessee	Kimberly Clark	0
Munising	Michigan	Kimberly Clark	0
Neenah	Wisconsin	Kimberly Clark	0
New Bern	North Carolina	Weyerhaeuser	0
Oswego	New York	Hammermill	0
Palatka	Florida	Georgia Pacific	0
Plymouth	North Carolina	Weyerhaeuser	0
St. Helens	Oregon	Boise Cascade	0
Sartell	Massachusetts	St. Regis	0

The disparity of results between the mills that have effectively eliminated TSS as a problem and those that appear to be ignoring it is glaring. Why were Crown Zellerbach's mills polluting at such a high rate while Kimberly Clark's mills were producing no TSS discharge? It is doubtful that differences in products or in the size and nature of the waterways caused these disparate results. Management must decide whether to pollute.

In Table 11.8 the companies are scored and ranked based on how well they abated TSS emissions. The results show the improvement in reducing TSS over the period and highlight the disparity between companies. In terms of the relative rankings, Champion improved the most, going from 5th in 1988 to 1st in 1996. St. Regis also improved from 4th in 1988 to 2nd in 1996. Most of the other companies either stayed in about the same place or suffered a deterioration in ranking. Of the companies that were among the worst polluters, Crown Zellerbach (which went from 12th to 15th) and Georgia Pacific (which went from 6th to 12th) had the largest change in relative rankings. Mead, Hammermill and Westvaco also suffered a deterioration in rankings, but they were among the companies that emitted the least TSS in 1988.

Table 11.8. Companies Ranked on TSS Emissions 1988, 1993 and 1996

	1988		1993		1996	
Company	**Score**	**Rank**	**Score**	**Rank**	**Score**	**Rank**
Boise Cascade	26.76	9	19.81	10	11.54	8
Champion	14.84	5	4.09	1.07	1	
Crown Zellerbach	45.41	12	36.10	14	100	15
Georgia Pacific	15.78	6	36.31	15	21.48	12
Great N. Nekoosa	17.72	7	9.12	7	4.35	3
Hammermill	1.38	1	8.86	5	5.99	5
Int'l. Paper	62.73	13	31.18	13	46.32	14
Kimberly Clark			8.94	6	5.90	4
Mead	4.72	3	20.29	11	14.72	10
Potlatch	23.52	8	14.24	9	12.74	9
St. Regis	8.87	4	2.80	1	2.11	2
Scott	36.08	11	28.09	12	21.60	13

	1988		1993		1996	
Company	Score	Rank	Score	Rank	Score	Rank
Union Camp			5.47	4	8.12	6
Westvaco	1.47	2	4.44	3	8.72	7
Weyerhaeuser	27.13	10	13.60	8	17.92	11

pH

In terms of pollution control, the greatest success that the mills overall have shown is in pH. Most of the mills included in the study were able to discharge water that had a pH between 6 and 8.5. As indicated by Table 11.9, there has been a tremendous improvement from 1988 to 1996. In 1988 Champion was the only company that had no pH problem at any of its mills. By 1996 six additional companies had no pH problems. Furthermore, Boise Cascade, Westvaco and Great Northern Nekoosa each had one mill with a minor pH problem in 1996. Based on the information presented in this table, it would appear that all the firms should be capable of eliminating whatever pH problem they exhibit.

Table 11.9. Comparison of pH Scores by Company

		pH Score	
Boise Cascade	.009	.202	.005
Champion	0	0	0
Crown Zellerbach	.182	.204	.123
Georgia Pacific	2.678	0	0
Great N. Nekoosa	.061	2.030	.055
Hammermill	.611	.090	0
Int'l. Paper	.099	.065	.063
Kimberly Clark		.038	.120
Mead	.125	.010	0
Potlatch	.483	0	0
St. Regis	.615	.232	.144
Scott	1.489	.514	.100
Union Camp		0	0
Westvaco	.413	0	.008
Weyerhaeuser	1.165	0	0

Despite the successful pH pollution control exhibited by the companies as a group, a number of mills were still having problems with too high or too low a pH in 1996. Table 11.10 lists the mills that were still having problems controlling their pH. Scott had the most mills with a pH outside the acceptable range. Based on these data, only 2 mills had real problems in 1996: Fort

Table 11.10. Mills with pH outside the Acceptable Range, 1996

Mill	Location	Company	pH (1996)
Fort Edward	New York	Scott	.46, 10.33
Spotswood	New Jersey	Kimberly Clark	11
Corinth	New York	Int'l Paper	5.07
Winslow	Maine	Scott	5.2
Bucksport	Maine	St. Regis	5.2
Port Angeles	Washington	Crown Zellerbach	5.45
Camas	Washington	Crown Zellerbach	5.5
Marinette	Wisconsin	Scott	5.8
Port Edwards	Wisconsin	Great N. Nekoosa	5.8
Luke	Massachusetts	Westvaco	8.67
Steilacoom	Washington	Boise Cascade	5.9

Edward and Spotswood, both located in the Northeast. Furthermore, the 5 worst mills with pH problems were all located in the Northeast.

Since only 11 of the mills had any problem at all with pH in 1996, and realistically, at most, only 7 had more than a slight problem, it seems fair to conclude that the mills have solved the problem of pollution caused by a too high or too low pH.

In Table 11.11, companies are ranked on their pH performance for the years of study. In comparing the scores of 1988 and 1996, it is clear that there has been a great improvement by the companies over this period. For example, Boise Cascade, with a score of .45 in 1988, ranked 2nd. In 1996 the company had a score of. (an improvement of about 42%) and ranked 8th.

The highest score in 1996, 10.4 by Scott, would have given it a rank of 7th in 1988. Furthermore, six companies had no pH problems in 1986 compared to only one company in 1988.

Table 11.11. Companies Ranked on pH, 1988, 1993 and 1996

	1988		1993		1996	
Company	**Score**	**Rank**	**Score**	**Rank**	**Score**	**Rank**
Boise Cascade	.45	2	10	11	.26	8
Champion	0	1	0	1	0	1
Crown Zellerbach	15.1	7	10.74	12	6.09	12
Georgia Pacific	50.30	13	0	1	0	1
Great N. Nekoosa	1.19	3	100	15	2.72	10
Hammermill	5.10	5	4.46	9	0	1
Int'l. Paper	5.75	6	3.32	8	3.12	11
Kimberly Clark			4.55	10	6.44	13
Mead	2.18	4	.50	7	0	1
Potlatch	23.86	10	0	1	0	1
St. Regis	19.65	9	11.49	13	7.13	14
Scott	38.61	12	20.30	14	10.40	15
Union Camp			0	1	0	1
Westvaco	16.19	8	0	1	.40	9
Weyerhaeuser	37.67	11	0	1	1	1

The one troubling part of this analysis is that some of the mills still did not have pH totally under control. If most mills and companies can deal successfully with maintaining an acceptable range of pH, there is little excuse for not achieving this. Whether the problem is one of lax standards by the states or aberrant mill management, it should be brought under control.

Companies' Overall Pollution Performance

Most of the companies included in the study improved their overall water pollution performance over the study years. As is indicated by Table 11.12, only two companies had a deterioration in their overall performance: Mead and Hammermill. These companies, however, were among the better companies in controlling pollution in the years of the study. In terms of ranking, three companies improved the most: Champion, St. Regis and Great Northern Nekoosa, each of which was among the best for each year. The most radical improvements in terms of overall score were for the companies that were among the highest-ranked firms in 1996: Weyerhaeuser, Georgia Pacific, Scott and International Paper. Despite International Paper's major improvement, along with Crown Zellerbach, it had by far the worst overall performance in 1996.

What the table clearly indicates is that some companies were much more successful than others in controlling water pollution. Since there is such disparity in results, and since most of the mills were in compliance with the water pollution standards, it appears that this method of reducing water pollution is a function of management's commitment to the problem. However, depending on the whims of management to achieve a desired level of pollution cannot be the most optimal way of achieving the goals of the Federal Water Pollution Control Amendments.

Table 11.12. Water Pollution Performance Score and Ranking

	1988		1993		1996	
Company	Score	Rank	Score	Rank	Score	Rank
Champion	4.86	1	8.25	1	33.38	4
St. Regis	14.68	2	21.41	3	40.48	5
Great N. Nekoosa	15.65	3	128.76	15	69.76	7
Westvaco	18.08	4	12.01	2	24.72	3
Potlatch	23.81	5	35.86	5	103.38	8
Union Camp	25.01	6	25.28	4		
Mead	25.46	7	38.55	6	12.92	1
Boise Cascade	29.71	8	55.30	10	57.44	6
Hammermill	33.59	9	44.73	7	17.60	2
Kimberly Clark	39.06	10	46.23	8		
Weyerhaeuser	49.90	11	47.16	9	121.27	10
Georgia Pacific	53.83	12	86.91	12	141.92	11
Scott	59.69	13	101.01	13	112.22	9
Int'l. Paper	106.02	14	70.98	11	168.10	13
Crown Zellerbach	144.10	15	123.07	14	160.51	12

Based on the findings in this study, it is apparent that a new approach to reducing water pollution is needed. Although the current approach has led to a reduction in harmful discharges, after 20 years one would expect less disparate results and much greater success. Since the federal government has chosen to try a technique of emissions trading to deal with air pollution, perhaps a similar approach can be tried with water pollution. This seems especially feasible since point source data for water pollution is available and the system for determining the amount of allowable water pollution has been in place for almost 20 years.

□□□

12

Measurement of Pollution Performance

Introduction

The most perplexing issue relating to the evaluation of pollution performance is to develop appropriate tools and techniques to measure the firm's pollution performance. It is important that the measurement techniques are suitable to the nature of pollution data and capable of providing reliable information on the firm's pollution abatement activities.

First we will discuss the measurement techniques utilized by earlier studies. Thereafter, we will explain the measurement techniques used in the study on which this book is based. We will also briefly present the pollution problem in the pulp and paper firms as well as in electric utilities and identify important pollutants for which the measurement techniques are being developed.

Measurement Techniques Used

The existing major studies dealing with the measurement of pollution information have been conducted under the auspices of the Council on Economic Priorities (CEP). These studies have covered firms from the following four industries: pulp and paper, steel, electric utilities, and oil refining. We will evaluate the measurement methodology used by the first three studies. Therefore, the CEP study for oil refining firms will not be included in the review of methodology used by the existing studies.

The main purpose of the study on pollution in the pulp and paper industry conducted by Allen, Kaufman and Underwood "was to ascertain what the 24 leading producers of pulp and paper in the United States had actually accomplished in controlling pollution at their pulp mills." That study also evaluated the degree of commitment of individual company management for pollution abatement and estimated the amount spent by each company to achieve the state-of-the-art control at each of their mills.

The study by Allen et al. covered 130 pulp mills operated by the 24 largest pulp and paper companies. All mills were involved in producing "virgin pulp" by different processes. Some mills used recycled wastepaper as raw material, but mills where only recycled paper was used were excluded from the study. The "virgin" pulping facilities were especially focused because their operations were considered to be the most polluting of all paper production processes. The mills covered by the study were scattered across the country. The study focused on the kind of pollution problem that existed at each location, and on the steps to be taken and expenditures needed to correct the problems. It also evaluated the specific control equipment installed and its effectiveness.

Data for the study were collected through interviews, visits and correspondence with each of the companies' executives who was charged with managing air and water pollution control efforts. In order to have a better understanding of the range of pollution problems, the researchers also corresponded and interviewed the representatives of air and water pollution regulatory agencies in each relevant state. Additionally, views of the environmental groups and/or citizens in and around the mill towns were obtained through interviews and all federal and state pollution abatement hearings for the past five years were reviewed.

On the basis of information obtained from different sources, the researchers provided a pollution overview of all companies and summarized the company's pollution control record. The findings also included estimates of what the company may have to spend to control pollution. Additionally, mill-by-mill analyses were provided, which included the company overview, the history, type and capacity of the plants. Each mill was rated in a chart form on the basis of the efficiency of its pollution control system in six different categories. The specific types of control equipment installed with its rated or active efficiency were given for each category. The overall evaluation of the ability of the installed equipment to meet the state-of-the-art criteria was indicated as adequate or not adequate.

In the summary of findings, the study indicated that the 130 mills surveyed produced 88,904 tons of pulp daily using at least 27 billion gallons of water. The obnoxious odor was adequately controlled at only 32 of the sites, and 46 sites completely failed to provide adequate odor control. As far as water pollution is concerned, two-thirds of the total 2.7 billion gallons used daily failed to satisfy the standards for primary and secondary treatment. Almost no water treatment was given to 895 million gallons of effluent (330% of the total) discharged every day. Another 930 million gallons (34%) were treated only for removal of heavy solid wastes. Full secondary treatment was provided for 868 million gallons (3%). The study concluded that 2 of the 24 companies surveyed had a record of sustained interest and effort toward achieving excellent pollution control at their mills.

The methodology used by the CEP study thus primarily consisted of observations, interviews and a review of records. The results of the study were descriptive in nature. No attempt was made to provide quantitative measures for mills' pollution performance. The methodology used in this book for measuring water pollution differs significantly from that used by Allen et al. In this book, quantitative measures for pollution emissions have been developed, and the methodology used for the development of these measures has been borrowed from the CEP steel industry study, which we modified to suit the water pollution emissions by pulp and paper firms.

The CEP study *Environmental Steel, Pollution in the Iron and Steel Industry*, which was conducted by Cannon, was based on particulate and sulfur dioxide emission data. The pollution data were obtained from the permit applications filed by the firms with the regional offices of the U.S. Army Corps of Engineers; duplicate copies of permits were, however, also available from the regional U.S. Environmental

Protection Agency offices. The study indicated that, approximately 28,000 Reform Act permit applications had been filed by U.S. corporations for their industrial water discharges, and the study covered all permit applications filed by 47 steel mills.

Each Reform Act application has been divided into two sections. The first section, in addition to name of the company, name and location of the mill, and number of employees, also contained information on the gross water intake and its sources, the amount of daily water usage within the mill, and the amount of discharge. The content and volume of the discharged water is described in Section II of permit applications, which contained information on each outfall pipe discharging the waste water.

The impact of each mill on its receiving waterways was determined in two ways. First, the net amount of each of 20 different water pollutants being added to the receiving waterways was determined, and second, the average increase in the concentration of each pollutant as it passed through a mill was calculated. The following formula was used to calculate the daily net addition of each pollutant (except for temperature and pH). The net addition was defined as the water pounds discharged over and above what was present in the intake water.

$$1 - \frac{a}{b} \times C = p \quad (1)$$

where a is the intake concentration of particular pollutant, b is the discharge concentration of the same pollutant, C is the total pounds of this pollution discharged daily.

For example, the calculation for the suspended solid was done as follows:

1 - 10/15 x 125 = 42 pounds of suspended solids added to the receiving water for outfall no. 1 of each day.

In cases where a company neglected to report the total pounds of a pollutant discharged daily (c), the net pounds figure was calculated as follows:

$$(b - a) \times 8.3 \times \mathrm{d} = \mathrm{p} \quad (2)$$

where a and b are the same as above, 8.3 is a conversion factor from parts per million to pounds per million gallons, d is the daily flow rate from the outfall in million gallons per day.

For example, if the hypothetical outfall no. 1 has a daily discharge of 1.0 million gallons of water, the net suspended solids addition will be:

(15 – 10) x 8.3 x 1.0 = 42 pounds of suspended solids added to the receiving body outfall no. 1 each day.

The total net pound of a pollutant discharged by each mill was obtained by adding the net pounds contributed by each outfall at the mill.

The study also determined a mill's pollution load in terms of increasing the concentration of each pollutant at any mill. This measure was considered to be important because it treated large and small mills equally. Additionally, from the health aspects of pollutants, it was more relevant because the health aspects are described in relation to the concentration level rather than pounds of pollutants in water. To determine concentration increase for any pollutant, the average intake concentration (C_i) and average discharge concentration (C_d) must first be calculated. The average intake concentration can be determined from the outfall data using the following formula:

$$\frac{d_1a_1 + d_2a_2 \ldots d_na_n}{D} = C_i \qquad (3)$$

where $a_1, a_2, \ldots$ an is intake concentration of the pollutant at outfalls no. 1, 2, n; $d_1, d_2 \ldots d_n$, is daily flow rate at outfalls no. 1, 2, n; n is the number of outfalls; D is the daily flow rates from all the outfalls at the mills.

Similarly, the average discharge concentration for any pollutant at any mill was determined using the following formula:

$$\frac{a_1b_1 + a_2b_2 \ldots d_na_n}{D} = C_d \qquad \ldots(4)$$

where $b_1, b_2 \ldots b_b$ is discharge concentrations of the pollutant at outfalls no. 1, 2 . . . n; $d_1, d_2 \ldots d_n$ are the same as in above formula.

The difference between the intake and discharge concentration provided information on whether the company increased the concentration of suspended solids. This difference represented the pollution emission by each pipe; the sum of pollution emissions by all pipes in a mill represented pollution by the mill, and the firm's pollution emission was obtained by adding pollution emissions by all mills in the firm. The research methodology used in the White study was superior to the methodology used in the earlier CEP study on electric utilities and provided a better base for a comparative analysis of pollution performance of electric utilities over different time periods.

In his study, White used the following three measures of air pollution: particulates, SO_2 and NO_x. The author calculated these measures for each of the fossil fuel-burning power plants of each company and presented a plantby-plant analysis for each company. Furthermore, each pollutant was measured for the company as a whole and then a comparative analysis of companies was conducted with regard to their emission levels. The methodology used by White has been adopted for the study on electric utilities on which this book is based. This methodology will, therefore, be described later in this chapter under the heading of "Measurement of Air Pollutants."

Measurement of Water Pollutants

The water pollution measurement techniques used in this book have been developed by modifying and refining the methodology used by Cannon. The methodology was modified to suit the nature of pollution data generated by pulp and paper firms. The raw pollution data for the measurement of pollution performance of pulp and paper firms were obtained from the pollution reports filed by the firms with the regional offices of the Environmental Protection Agency (EPA). The pollution reports filed by firms provided information on each pollutant for each individual pipe of their mills. The raw data for each pollutant were processed first to develop pollution scores for individual pollutants for each pipe and mill, and then these scores were aggregated to arrive at a pollution index for the firm as a whole.

Water Pollutants for Pulp and Paper Firms

In order to measure the extent of water pollution contributed by paper and pulp mills, the first important step is to identify major pollutants which are considered harmful and on which the firms are required to report. These pollutants will then form the basis for the development of a pollution index for pulp and paper firms. The three water pollutants analyzed are: BOD, TSS and pH. We will describe the methodology to measure each of these pollutants and to develop an overall pollution index.

BOD Pollution Measure

The BOD pollution score was calculated in three steps. First, the pollution score for each pipe was determined. Second, pipe pollution scores were aggregated to obtain the BOD pollution score for the mill. In the third step, the BOD pollution scores for the mills were used to obtain the pollution score for the firm.

BOD Pollution Score for Each Pipe. In order to measure the BOD pollution score for a pipe, we required intake and outflow of the pollutant for each pipe, and we also required concentration and weight of the pollutant.

The pollution reports filed by the firms contained information on BOD outflow as well as on the BOD weight. Additionally, some reports also contained information on BOD concentration. But unfortunately, most of these reports did not contain information on BOD intake. Research was conducted to determine the most appropriate value for BOD intake. Based on a report by McDuffie and Haney a value of 5 was considered to be the most appropriate.

The BOD pollution emission by a firm was termed as Ä BOD $_i$ which was calculated as a difference between the BOD outflow and BOD intake considering the BOD weight and BOD concentration. If values for BOD weight and BOD concentration were available from the EPA reports, Ä BOD was computed as follows:

$$\Delta BOD_{ijk} = \left| 1 - \frac{5}{BODC_{ijk}} \right| \times BODW_{ijk} \qquad \ldots(5)$$

where BOD_{ijk} = Pollution emission from pipe k in mill j of firm i; $BODC_{ijk}$ = BOD concentration from pipe k in mill j of firm i; $BODW_{ijk}$ = BOD weight from pipe k in mill j of firm i.

If the value for *BOD* weight was not available from the EPA reports, the pipe's water flow was used to calculate the BOD weight. The water flow value was multiplied with a factor of 8.3, the same factor suggested by McDuffie and Haney in their study.

$$BOD_{ijk} = [BODC_{ijk} - 5] \times 8.3 \times F_{ijk} \qquad \ldots(6)$$

where F_{ijk} represents an average waterflow for each pipe k in plant j of firm i.

The average value of waterflow was computed over the time period of the study. If the value of BOD_{ijk} was in the minus range, it was converted to 0, meaning that there was no BOD pollution emission by pipe K.

❑❑❑

13

Legal and Regulatory Framework in Groundwater Contamination

Introduction

Contamination has occurred in most regions of the country and will probably continue to occur. Once contamination is detected there is a choice as to whether to take action and, if so, what form that action should take. The type of actions vary with the type of contamination and its hydrogeologic and geographic setting. There are regulatory agencies at the federal, state, and local levels that have requirements that may prevent or correct many types of groundwater contamination. In consideration of the future quality of groundwater the question is whether or not to take action that would completely or partially prevent such contamination.

Importance of the Groundwater Resource

More than 50% of the population depends upon groundwater as its primary source of drinking water. Approximately 75% of American cities derive their supplies, in total or in part, from groundwater. The states vary in their dependence upon groundwater. Those in the arid West, where there are few permanent rivers, lakes, or streams, are much more reliant on groundwater than the states of the more humid East. In 1980, 88 billion gallons of groundwater were used in the United States per day, and 68% of this total was used for irrigation. The total use in 1950 was 34 billion gallons per day (bgd), and in 1970, 68 bgd.

Definition of Groundwater

Groundwater, as considered here, is water that occurs in saturated non-consolidated geologic material (sand or gravel) and in fractured and porous rock. These saturated strata are called aquifers. Groundwater comprises approximately 4% of the water contained in the hydrologic cycle, which consists of the movement of water between oceans and other bodies of surface water, the atmosphere, and the land. Groundwater is recharged by precipitation.

Types of Aquifers

The two main types of aquifers are unconfined and confined. An unconfined aquifer is not overlain by impermeable material, and precipitation may percolate down to the water table. Water in the aquifer is usually at atmospheric pressure. The amount of water in storage is dependent on seasonal cycles of natural recharge. A perched aquifer may occur where there is a limited layer, or lens, of impermeable material above the water table, forming a zone of saturation above it.

The second major category of aquifers is the confined aquifers, some of which are artesian. Confined aquifers are bounded top and bottom by layers of relatively impermeable geologic formations termed aquitards. Artesian aquifers contain water that is under pressure greater than atmospheric. Some have discrete and variable recharge areas where the geologic material of the aquifer forms an outcrop at the surface. A few aquifers have no recharge areas at all.

The areal extent of aquifers may be regional or local, and some, such as the Ogallala in the Midwest, may extend across several states. This fact is important when one comes to consider the institutional organization for the management and protection of groundwater.

Most aquifers occur at variable depths within 2,500 feet of the land surface. Estimates of water in storage within 2,500 feet of the 3 land surface within the conterminous United States range from 33 quadrillion gallons to 100 quadrillion gallons. It would be possible to drill a well with a yield sufficient for domestic use in almost every region of the country.

Factors Affecting the Natural Quality of Water

The natural quality of groundwater is affected by the geological characteristics and climatic conditions n particular, the temperature and rainfall of the region. When evaporation and transpiration exceed precipitation, as in the arid Southwest, the recharge of groundwater takes place only in "wet," or multicycle years. Some aquifers, such as the Ogallala, have little, if any, recharge. In the East, where precipitation usually exceeds evaporation, unconfined aquifers are easily recharged. The natural chemical content of groundwater is influenced by the type and depth of soils and by the subsurface geologic formations through which the groundwater passes. The leaching of the soil and rocks is influenced by the pH of the precipitation and by the metabolic products of subsurface organisms. In the aquifer the water reacts with the geologic material, usually increasing its content of total dissolved solids (TDS). U.S. Public Health Service drinking-water standards consider 500 parts per million (ppm) TDS the maximum permissible, although the Safe Drinking Water Act regulations consider water with up to 10,000 ppm TDS as potentially potable. The hardness of water reflects its calcium and magnesium content and may render the water unfit for domestic or industrial use. Water from some aquifers may be unusable due to high salinity or to the presence of naturally occurring toxic substances such as arsenic and radionuclides. Natural groundwater may be unfit for certain uses or be of the highest quality that can be used for drinking water with no pretreatment.

Relationships Among Groundwater Quality and Quantity, Surface Water, and Land Use

In the arid areas, where evaporation and transpiration can greatly exceed precipitation, the mineralization of groundwater due to leaching is a significant cause of contamination; evapotranspiration may further concentrate the salts. Pumping at greater than recharge rates can increase the degradation of groundwater. If pumping lowers the water level in the vicinity of coal mines, the exposed sulfur-bearing rocks may be subject to oxidation. Should the water levels increase, the oxidized sulfur-compounds may dissolve and increase the acidity of the water. In coastal areas, where there is a high demand for water, groundwater pumping may cause saltwater intrusion into potable aquifers.

Most of the low flow of streams is supplied by groundwater. If the aquifers are overpumped and recharge does not occur, the stream flow may cease during periods of drought. Moreover contamination of groundwater by contaminated streams may occur if a negative pressure develops in the aquifer from the use of groundwater.

The recharge of aquifers may be greatly reduced by paving the land surface in recharge zones or by altering the vegetation cover so that precipitation runs off instead of infiltrating. Siting of contaminating activities in recharge areas without careful consideration of practices that could eliminate or reduce contamination may cause a threat to the quality of groundwater and hence its usability. Thus, the relation between groundwater quantity and quality, between groundwater and surface water, and between groundwater and land use patterns must be taken into consideration in any protection and management strategy.

Sources of Groundwater Contamination

There are three main mechanisms by which the chemical composition of groundwater may be changed: by natural processes, by man's waste-disposal practices such as those for sanitary wastes, liquid industrial wastes, solid wastes, and radioactive wastes, and by spills, leaks, and agricultural activities and other sources unrelated to disposal.

Types of Groundwater Contamination

The types of contaminants which emanate from anthropogenic sources are extremely varied. They range from simple inorganic ions such as the nitrate from septic tanks, feedlot wastes, and the use of fertilizer, chloride from highway deicing salts, saltwater intrusion, and certain industrial processes, heavy metal ions from many different industrial processes such as plating works, to complex synthetic organic compounds resulting from industrial and manufacturing processes and from the use of pesticides and household cleaning fluids, e.g., trichloroethylene. Compounds of low solubility may yield solutions that are toxic or offensive. The chemical composition of wastes deposited in landfills or surface impoundments is often known; nevertheless, when the constituents of such wastes interact, new compounds may be formed. Many industrial waste-disposal practices now include the stabilization of wastes, thereby rendering them less active chemically. It is still possible, however, that leachate production may transform some of the constituents.

Attenuation and Movement of Contaminants in Groundwater

A contaminant usually moves from the surface of the land, percolating downward through the non-saturated zone into a semisaturated or saturated zone and into the aquifer. Contaminant movement through the subsurface is determined by its solubility in water, its density, and the various attenuation processes that may occur below the surface. The attenuation of contaminants may take place by biological activity, dilution, volatilization, mechanical filtration, precipitation, buffering, neutralization, ion exchange, radioactive decay, hydrolysis, complexation, and adsorption. These processes may reduce or increase the severity or toxicity of the contaminant, and may also alter its rate of movement. Different soils, and soil conditions have different capacities for attenuation, and affect various chemicals to different degrees. At present these attenuation processes are being researched and our knowledge is limited. Once in the aquifer the contaminant movement is mainly governed by advection, dispersion, and diffusion; the flow pattern is often laminar. Because contaminants slowly mix with water and turbulence is lacking, a well defined plume may be formed until contaminants completely diffuse into the aquifer. The movement of contaminants is usually but not always with the direction of flow. The miscibility of contaminants and their specific gravity are also important characteristics in determining movement. These factors determine whether contaminants float on top of the water table or sink.

Risks Posed by Different Types of Contaminants

The degree of risk posed by contaminants varies according to many factors. These include the volume and toxicity of the contaminant, its concentration in the aquifer, its persistence in the environment, and the degree of human and environmental exposure to the contaminant. In addition, the number of persons affected, or likely to be affected, over time and the percentage of available groundwater both locally and regionally should be taken into consideration. If the contaminants in the groundwater exceed the standards set for drinking water by the federal government, for example, then the water is hazardous for the use for which it was designated under the standards. These standards include, however, only a limited number of chemicals, and thus they do not necessarily protect humans or the environment against either the short-term or the long-term effects of every contaminant that might be found in groundwater. Drinking water standards are designed to protect human health; they may not protect other organisms.

One of the main criteria usually used in judging risk is the contaminant's effect upon public health. The epidemiological evidence for an association between chronic disease and the use of contaminated groundwater is inconclusive but suggestive of a link with certain diseases. More is known about the acute effects, with regard to disease caused both by pathogenic microorganisms and by chemicals, of ingesting contaminated groundwater. Some of the synthetic organic chemicals that have been detected in groundwater have not been tested for their potential to affect human health, and thus we do not know whether they pose a hazard.

The effects upon the environment of contaminated groundwater are even less well documented than its effects on human health, although it is well established that certain contaminants may cause stress to vegetation or even its destruction, and that they probably adversely affect animals. For example, water for irrigation should be relatively free of boron and chloride, both of which have a deleterious effect upon plant growth. Many plants selectively accumulate heavy-metal ions from water. Water for human use should ideally be of the highest quality, but livestock may also be particularly sensitive to contaminants at concentrations that are not injurious to humans.

Another way to assess the potential danger of contaminated groundwater is to gauge the persistence of its deleterious effects on the environment. For example, cadmium, which is toxic and believed to be carcinogenic, has an infinite half-life and is not 7 degraded. Some inorganic and organic compounds may be degraded by biological action in the soil zone, so that only harmless end products reach the aquifer. All of the heavy-metal ions and some synthetic organic chemicals associated with manufacturing and service industries would probably not be degraded.

Geographical Extent of the Problem

No comprehensive national survey of groundwater contamination has been undertaken, for the most part because it would be extremely expensive and time-consuming. There have been many federal and state-run surveys, conducted on a regional and local level, on ambient-water quality, drinking-water supplies, and potential sources of contamination. Incidents of contamination have been reported from every state, and the incidents are beginning to be detected with increasing frequency. From the assessments listed above, it may be concluded that problems indigenous to one area may not occur in another, but that several sources of groundwater contamination occur at a high or moderate degree of severity in each area studied. The four types of pollutants most commonly reported (heavy metals, organics, microorganisms, and nitrates) are probably more of a reflection of the prevailing monitoring practices at the time the surveys were made than they are of the actual incidence of the contaminants that most threaten the quality of groundwater.

For this report we selected nineteen states for which we had case histories of incidents of contamination (AZ, CA, CT, FL, ID, IL, MA, MT, ND, NE, NJ, NM, OR, PA, RI, SC, TX, VT, and WA). These states represent areas of differing levels of industrialization, agricultural activity, climatic conditions, and population density.

Analysis of these states indicates that in many of them industrial and municipal wastes were among the most frequently reported sources of contamination. The industrial Northeast predictably had more problems associated with industrial wastes, petroleum products, and landfill leachate. California, Florida, Idaho, Nebraska, and Texas reported problems arising from agricultural practices, whereas California, New Mexico, and Texas reported problems with the disposal of oil-field brines. These summaries 8 include, of course, only the known incidents of contamination; a comprehensive national survey might well uncover other important sources of contamination, or different frequencies from the same sources. The problems encountered to date vary regionally and depend upon the type and degree of industrial and agricultural activities carried out and the population density. The known incidents vary from a low to a high frequency of occurrence of contamination, and they fall into geographical patterns.

Severity of the Problem

In order to determine the overall magnitude of the problem, it is necessary to assess the severity of the contamination of groundwater and this can be undertaken in several ways. If the concentration of contaminants exceeds the numerical standards set for a particular use, as, for example, for drinking water, the problem could be called severe. The severity depends on the population affected by well closings, the nationwide percentage of aquifers affected, and the percentage of the known reserves of groundwater that are contaminated. The degree of hazard posed by the contaminants for example, their toxicity enters into the evaluation, as does the cost and feasibility of finding an alternative water supply. Many of the data required to assess the severity of the problem by these methods are unavailable.

Since no systematic national survey of the contamination of groundwater has been carried out, the only estimates of the order of magnitude of the amount of groundwater currently contaminated rely upon surveys of the numbers of certain types of potential sources of contamination. EPA did such an estimate for certain selected point sources, such as surface impoundments and landfills, which it considered the most important sources, and also for secondary sources, such as subsurface disposal systems and petroleum exploration and mining. The agency concluded that up to 1% by area of usable surface aquifers near the land surface may be contaminated by these sources. Lehr estimated that up to 2% could be contaminated. It should be pointed out that these are nationwide estimates for the entire United States. These estimates did not include contamination from non-point sources, such as agricultural activities. In certain areas, especially those with a high population density, a much higher percentage of the groundwater may be contaminated.

Monitoring for Groundwater Contamination

Monitoring groundwater quality and aquifer conditions can bring to light contamination before it becomes a significant problem. The appropriate design of monitoring system depends upon the hydrological, geological, and soil characteristics of the site, pollution sources, population density, and climate of the region. The four main types of monitoring systems are those which measure quality in relation to numerical standards, source monitoring for effluent quality and quantity, case-preparation monitoring for enforcement-action data, and research monitoring.

In order that the samples of water withdrawn for analysis are not altered chemically, well-casing materials and sampling procedures must be carefully chosen. Detailed knowledge of groundwater flowpatterns, natural water quality, and the attenuation capacities of the soil and geologic strata would enhance the siting of monitoring facilities so that they produce the most useful data with the minimum of damage and expense. The chemicals analyzed should be the best indicators for the suspected type of groundwater contamination under investigation. They should include organic chemicals from natural and man-made origins as well as inorganic ions, such as chloride, nitrate, and heavy metals. The sophistication of present-day analytical methods may have outstripped our ability to interpret what they reveal and our ability to determine the significance of low-concentrations of contaminants on the environment and on public health.

Measures for Protecting Groundwater

The protection of groundwater from contamination may be costeffective for valuable aquifers. Many federal and state regulations can be used for this purpose, as summarized below, although the majority of regulations were not specifically designed with groundwater protection as their primary goal. The most efficient way of preventing important aquifers from becoming contaminated is by protecting their recharge areas. For most confined aquifers it would be necessary to prohibit the siting of contaminating activities in specified recharge areas or to ensure that potentially contaminating activities use adequate methods to prevent groundwater degradation. For unconfined aquifers this might be difficult because of their extensive recharge areas. Protection might entail careful regional and local planning since recharge areas may cross county or state boundaries. Other protective measures can be taken, including the careful restriction of groundwater pumping in areas where salt water can intrude into potable aquifers, the capping and sealing of dry boreholes that traverse one or more aquifers and offer a conduit for contaminants, and the continued stringent supervision of the drilling and maintenance of injection wells and water wells.

Aquifer Rehabilitation

Remedial action for contaminated aquifers is complex, timeconsuming, and extremely expensive. It is feasible under certain circumstances, assuming no additional contamination occurs, and the choice of remedial action determined upon is dictated by the time period during which contamination has taken place, the type and behavior of the contaminants, and the hydrogeology of the site. Often it is more cost-effective to locate a new source of water than to attempt treatment. Contamination that occurs in shallow plumes in unconsolidated rock may be controlled by the excavation and removal of the plume, once the problem of the final disposal of the contaminated material has been solved. There are numerous remedial alternatives: in situ treatment methods, withdrawal treatment techniques, and final disposal. In-situ techniques include the detoxification, stabilization, and immobilization of contaminants and require the use of biological cultures, chemical reactants, or sealants. The aim in each case is to detoxify or stabilize the contaminants or to form an impervious barrier around the plume in order to prevent its migration. Methods for withdrawal, treatment, and final disposal may use collection or withdrawal wells, subsurface gravity collection-drains, impervious grout-curtains, and cut-off trenches. Treatment of the water removed could be by reverse osmosis, ultrafiltration, ion-exchange resins, wet-air oxidation, ozonation and ultraviolet radiation, coagulation and precipitation, aerobic biological treatment, or the use of activated charcoal filters.

Costs, ranging from several thousands to several billions of dollars, have to be estimated on a case-by-case basis and are determined by many factors specific to the site. For many problems 11 pertaining to groundwater pollution, the prevention of contamination is far better than curative action.

Ground-Water Protection Strategy

The EPA set forth a Ground-Water Protection Strategy which addressed four major objectives needed for effective groundwater protection: the need to strengthen state groundwater programs; the need to address groundwater contamination problems of national concern; the need to create a policy framework for guiding EPA programs; and the need to strengthen internal groundwater organization. In order to strengthen state groundwater programs EPA has provided states with funds to: establish or begin state strategies; support necessary program development and planning; create needed data systems; assess legal and institutional impediments to comprehensive state management; and develop state regulatory programs such as permitting and classification. To address groundwater contamination problems of national concern EPA has undertaken several reviews of a number of groundwater contamination sources that have appeared to be a widespread problem, for example leaking underground storage tanks, land disposal facilities, and non-point source pollution. In the strategy, EPA also recognized that in order to effectively protect groundwater, guidelines were greatly needed for the assurance of consistency within EPA's groundwater protection programs. The strategy's framework is based on the rationale that not all groundwater is of equal value, and the extent of protection should be based on its vulnerability to contamination and its respective value. Lastly, in order to strengthen internal groundwater organization, EPA has established the Office of Groundwater Protection to oversee the implementation of this strategy.

Classification of Aquifers

The classification of aquifers is a management tool that is in use in some states and under consideration in others. It establishes waterquality goals for each aquifer and identifies the standards or controls necessary to assure that those goals are met. A policy that chooses selectively to protect different aquifers at different levels of quality is dependent upon a classification system. Because 12 hydrogeologic conditions, water quality, and present and future groundwater use vary from state to state, there is no aquifer classification system in use that could serve as a model for other states wishing to develop such a system. The policies of preserving and restoring aquifers for drinking water and of prohibiting the degradation of existing water quality would not require classification systems.

The benefits provided by using a system of aquifer classification include legal protection for valuable aquifers, a basis for siting potentially contaminating activities in low-risk areas, and guidance for planning programs which would protect the quality of water at all levels of government. The deterrents to establishing a classification system include the difficulty of delineating aquifer boundaries, the difficulty of gaining public acceptance for degradation zones or waste-receiving zones, and the difficulty of projecting groundwater use, needs, and availability for the future.

Federal Regulations Applicable to Groundwater

There is no single statute specifically aimed at protecting groundwater, there are many programs that provide some measure of protection. The most significant federal statutes affecting groundwater are the Resource Conservation and Recovery Act (RCRA), the Safe Drinking Water Act (SDWA), the Clean Water Act (CWA), and the Comprehensive Environmental Response, Compensation and Liability Act (CERCLA or "Superfund Law"). Statutes less directly related to groundwater include the Toxic Substances Control Act (TSCA), the Surface Mining Control and Reclamation Act (SMCRA), the Federal Insecticide, Fungicide and Rodenticide Act (FIFRA), and the National Environmental Policy Act (NEPA).

The Resource Conservation and Recovery Act (RCRA) along with the Hazardous and Solid Waste Amendments (HSWA) to RCRA, provides a comprehensive framework for the management of municipal and industrial solid and hazardous wastes. RCRA requires owners and operators of facilities that treat, store, or dispose of hazardous wastes to obtain RCRA permits and meet groundwater standards at facility boundaries. It aims to protect groundwater affected by facility operations and correct past contamination that occurred at the sites.

The Safe Drinking Water Act contains programs designed to provide for sanitary drinking water supplies, protect sole-source aquifers, and establish a program to control the underground injection of wastes. The amendments to the Safe 13 Drinking Water Act contain provisions that require states to establish "Wellhead Protection Areas" around public drinking water well fields. The Clean Water Act has limited statutory authority to implement groundwater protection measures. Its primary goal is the reduction and control of discharge of pollutants into the nation's navigable water.

The Comprehensive Environmental Response, Compensation and Liability Act (CERCLA, or "Superfund Law") is essentially a groundwater remediation law. It enables the federal government to respond immediately to releases or threatened releases of hazardous substances into the environment which pose a threat of imminent and substantial danger to public health. Contained within the act is the National Contingency Plan (NCP), which outlines response action for the release of hazardous substances and oil spills. The act also includes criteria for determining priorities for remedial actions among these releases in the Hazardous Ranking System (HRS). The HRS "score" is then used by EPA to determine whether sites should be included in the National Priority List. Many of the statutes outlined above contain broad imminenthazard provisions that enable EPA to take immediate action to restrain activities posing a threat to any feature of the environment, including groundwater.

State and Local Groundwater Policies and Management Strategies

Within the last decade states have made large-scale efforts to prevent, abate, and monitor groundwater pollution. In the past these efforts have unfortunately received less attention and fewer program resources than corresponding surface-water programs. More recently, however, the groundwater problem has received national attention, and a thrust toward groundwater protection is taking place within many state environmental institutions. Because of the variation in the natural quality of groundwater within each state, the regional characteristics of man-made contamination, and the industrial goals for development, each state has been given primary responsibility to establish, implement, and enforce groundwater protection programs. There are essentially three basic types of policies that states have chosen to implement (either alone or in combination): non-degradation, limited degradation, or differential protection. A non-degradation policy calls for the protection of groundwater at its existing quality. A limited degradation policy strives for minimum degradation of existing quality but allows for degradation up to a given standard. A differential protection policy, which is a use oriented policy, calls for different levels of protection based on groundwater characteristics, current needs, and anticipated uses. State regulations that affect the quality of groundwater fall into five broad categories:

- those dealing with particular sources of pollution, such as septic systems and waste disposal sites
- those establishing and implementing water quality standards for aquifer water, in which aquifers may be classified according to natural characteristics or to current or projected uses
- those regulating chemicals in effluent that may enter groundwater.
- those regulating well construction and use
- those regulating the use of land in areas overlying critical aquifer-recharge zones.

As with federal regulations, the state and local controls that effectively protect groundwater often are not designed for that purpose and have several other regulatory objectives.

Summary

- Groundwater is essential to the continuance of our present quality of life and for industry and agriculture.
- The quality of groundwater varies considerably because of natural and anthropogenic contamination.
- Estimates of the percentage of usable groundwater near the surface that is contaminated range from less than 1% to 2%,

which only takes into account contamination from certain point sources. In addition it should be emphasized that any estimate that is a nationwide average is not very meaningful because most areas would exceed or fall below the estimate. In addition, contamination is most likely to exceed these estimates in areas of high population density, where the number of persons affected may be large.

- Contamination incidents have been reported from every state in the nation.
- Sources of contamination are varied and include both point and nonpoint sources and may be planned or inadvertent.
- The contaminants range in type from simple inorganic ions to complex synthetic organic chemicals.
- The types of contamination vary from one region of the country to another and are influenced by climate, population density, intensity of industrial and agricultural activities, the hydrogeology of the region, and the status and enforcement of federal and state regulations that can be used to protect groundwater.
- The methods for assessing the severity of the problem of groundwater contamination are varied and include its effects on the environment and on human health, the margin by which water standards are exceeded for a particular use, the percentage of usable groundwater contaminated, and the ease and cost of finding an alternative water supply.
- The risks posed to humans and the environment vary according to the volume and toxicity of the contaminants, their concentration and persistence in the aquifer, and the degree of exposure to them.
- The impact of contamination on humans may be measured by well closings and illness caused by chemical or pathogenic contamination.
- The lack of comprehensive national surveys of the extent of groundwater contamination and the fact that few of the contaminants have been tested for their effects upon human health, make it difficult to assess, in quantitative terms, the national risk of using contaminated groundwater.

- Aquifer rehabilitation is feasible in certain cases but is expensive and time-consuming, with no guarantee of complete success.
- Aquifer classification based on water quality and present and projected use might be a tool for groundwater management and protection.
- Groundwater management must recognize the relation among land use, surface water, and groundwater.
- The prevention of the contamination of useful, potable aquifers is far superior to subsequent curative efforts. This can, in part, be achieved by regional planning and the enforcement of federal and state regulations designed to 16 protect groundwater.

❑❑❑

14

FUTURE APPROACHES FOR DIMINISHING WATER POLLUTION

Introduction

The instances of water quality management chosen for discussion in this chapter are those that appear to be the most highly developed regional approaches. In some cases, the laws are so new that there is little to report but good intentions. In others, there is a long history of activity on a regional basis. Obviously, these two situations have to be assessed somewhat differently. Many factors other than the "inherent merit" of legal and institutional arrangements will influence outcomes. Despite the limitations imposed by the difficulty of making comparisons at this time, we feel that the discussion will be useful. It is clear from legal developments and from professional and lay debate that we in the United States are trying to find our way toward suitable institutions for dealing with the water quality problem. In this regard, it is useful to review the experience of other areas, some of which have long had to face problems of an intensity we are just beginning to meet.

Regional authorities have also been established in Belgium, in Holland, and in Ontario, Canada, and a current research project points toward a regional approach in Upper Silesia in Poland. In the United States the Ohio River Valley Water Sanitation Commission has a record of some twenty years of productive effort.

At the end of the chapter we list some capabilities which a regional water quality management agency would seem to need to achieve economic efficiency. These provide us with a rather loose but useful framework for comparative discussion of the cases.

Ruhr Area Genossenschaften

The *Genossenschaften*, as described in the preceding chapter, have implemented a wide array of measures to improve water quality. These include treatment in large collective plants, waste treatment at individual municipal and industrial outfalls, mechanical reaeration of streams, retention ponds, diversion of a stream through a treatment plant, lowflow augmentation by releases from upland reservoirs, and specialized use of certain streams or stretches of streams. The costs of these facilities attributable to water quality management have been met largely from charges levied upon all waste dischargers according to the quality and quantity of waste they discharge whether it goes directly to treatment plants or into the river.

Historical Background. Critical conditions in the Emscher area at the turn of the century led to the formation of the *Emschergenossenschaft*-the first of the *Genossenschaften*, and the one on which the others were modeled. The naturally poor drainage situation in the Emscher valley became steadily worse after 1860 when coal mining began moving into the area, followed by steel, steel-finishing, and chemical industries. Subsidence in the mining areas soon created depressions that filled with the waste of industries and municipalities, and efforts by individual mining enterprises and small associations had little effect on over-all drainage conditions. A regional drainage plan that might have improved the situation was proposed in 1883, but abandoned when no agreement could be reached on membership arrangements.

Population in the area rose from 350,0000 in 1970 to over 1,250,0000 by 2000; industrial expansion accelerated; and the increasing waste discharges overburdened the rivers and streams. Individual communities and industrial enterprises began to pump water from the Ruhr watershed, and court cases involving claims of damages resulting from upstream waste discharges became frequent. Despite a few municipal treatment plants, sanitary conditions became worse, and when a typhoid epidemic swept the country in 1901 the Emscher area was particularly hard hit.

Towards the end of the nineteenth century it became obvious that forbidding mines, mills, or municipalities to discharge their wastewater into the rivers was a futile approach, and that substantial improvement could be obtained only by unified regulation of the Emscher from its headwaters to its mouth. In 1999 various industrial and municipal officials met in Bochum and appointed a commission, made up of representatives of municipalities, industries, and members of the scientific and technical community, to devise a technical plan for the area. The commission was also confronted with the question of how to form an organization that could successfully carry through the needed works. There was no organization that could be taken as a model, and the fact that the Emscher area was divided into some 200-odd political jurisdictions created additional problems. The commission was convinced that the plan could succeed only if a feeling of community could be established among the involved groups. This, in their view, required a self-governing organization with a carefully structured political representation. A higher unit of government would have to assume some supervisory responsibility, but it was argued that a continuing self-governing management agency would be the best means for economical operation; it would be in a position to adapt to the changing conditions of a dynamic industrial area and would have a close association with the technical and economic resources of the region. At this early stage, the need for technical competence and research work was stressed.

A question of great importance was who the members of the association should be. It was decided that the membership should consist of all dischargers of wastewater and those benefiting from the regulation of the stream. Not surprisingly, the matter of distribution of costs was prominently discussed by the commission, and the principle was established that costs should be distributed on the basis of the benefits received from the operation of the organization and the costs made necessary by the activities of the membership. In the Emscher, the most important costs were those related to subsidence in coal mining areas and those imposed by the discharge of wastewaters.

The proposed law passed the Rhennish and Westphalian Landtags and in 1904 was signed by the King of Prussia. The law sets out the duties of the *Emschergenossenschaft* in general terms as the regulation of the stream and treatment of wastewaters in the area. The statute indicates that this is to be done in accordance with a general plan, but it also provides for adaptation to changing circumstances.

The *Emschergenossenschaft* proved so effective that the other *Genossenschaften* were patterned after it, although none of them faced such critical problems at the time of their establishment. The organizational and membership arrangements are not precisely the same for all the associations, but on the whole the members fall into three groups: mines, industrial and business establishments; railroads and other productive facilities; and communities.

Present Organizational Arrangements. Because the *Genossenschaften* are organized along such similar lines, much of the following description of the *Ruhrverband-Ruhrtalsperrenverein* can be taken as generally descriptive of them all.

The basic political power in the Ruhrverband lies in the governing board which is made up of: (1) the owners of business and industrial establishments and other facilities that lie in the Ruhrverband area and that contribute to water quality deterioration in the Ruhr or its tributaries, and those who benefit from the activities of the Rhurverband to the extent that they make a specified minimum financial contribution to its activities; (2) communities within the Ruhrverband area; and (3) the *Ruhrtalsperrenverein* as a representative of the waterworks and other water withdrawal facilities.

The political organs are the assemblies and the board of directors. The assembly members (about 1,500) elect the board of directors, approve or disapprove the plan for water quality management, approve or disapprove the assessment of charges, and decide upon the basic method for calculating the level of charges. The assembly reaches its conclusions on the basis of absolute majority with the number of votes cast by each member being dependent upon the amount of financial contribution.

The board of directors consists of a chairman, his deputy, and seven additional members all elected by the assembly from among its membership. The business establishments, the communities, and the *Ruhrtaisperrenverein* must all be represented on the board of the Ruhrverband.

Since the board members serve only part-time and are not necessarily skilled in technical water resource matters, they appoint a staff director to manage the program. The staff of the *Ruhrverband-Ruhrtalsperrenverein* now comprises about 450 professionals and 400 other workers. In the course of its roughly fifty-year history, the Ruhrverband has built over 100 waste treatment plants on the Ruhr and its tributaries. The *Ruhrtalsperrenverein* has built and/or manages seven large upland reservoirs. The Ruhrverband also operates waterpower facilities and a series of shallow oxidation lakes in the Ruhr itself. The costs of these activities have been considerable. Construction funds are obtained from the capital market, while the annual expenses for interest, operation, maintenance and replacement are met by a system of charges. The basic charge for the Ruhrverband quality management activities is contingent upon the amount and quality of the effluent discharged by the industries and municipalities in the region. The staff prepares lists of charges based on waste discharge information. The lists show the name of the member and an estimate of the charge he must pay. The member may appeal his charge to a commission established by the Ruhrverband or finally to the courts. The Minister of Food, Agriculture, and Forests of North

Rhine-Westphalia has general supervisory powers and must approve the lists, but his responsibility is limited to determining whether they follow the form prescribed in the laws creating the association. The assessments on the individual members established through the indicated procedure are public obligations and can be enforced by law as taxes. However, because of particularly good relationships between the *Genossenschaften* and their members, there has been little need for enforcement. This favorable experience, plus the public character of the obligations, has given the *Genossenschaften* such a strong credit rating that there is an international market for their bonds.

River Authorities in England and Wales

In England and Wales uniform water policies are now laid down by the central government and administered in large measure by regional authorities. The British contributed heavily to the development of the basic biological sewage treatment processes and over the years constructed many plants. Indeed, most urban households are connected to some form of treatment plant. However, overloading and failure to deal effectively with industrial wastes in many instances have left the country with severe water quality problems.

A survey of stream and estuary water quality showed that 73 per cent of total river mileage was "unpolluted or recovering from pollution," 15 per cent was of "doubtful quality," 6 per cent was of "poor quality," and 6 per cent was "mostly polluted." At first sight these figures seem encouraging. But they fail to reveal that the long stretches of rivers with water of good quality are mostly in upland areas, that many estuaries are grossly polluted, and that the water in the middle and lower reaches close to urban areas tends to be of doubtful quality at best. These facts make the picture look quite different, for it is in the highly populated areas that the rivers are becoming increasingly important for water supply and for recreation.

The idea of entrusting the management of water resources to comprehensive authorities with jurisdiction over a whole river basin or group of basins is not new in Britain. It has been put forward from time to time over the years by Royal Commissions and government committees as well as by professional groups. Moreover, English law had been moving toward river basin management for some time. Several river basins, notably the Thames, have long had conservancy boards empowered to deal with navigation and pollution control. Catchment Boards were set up in every drainage district in England and Wales to carry out land drainage and flood control works, and in some river basins there were agencies with authority to enforce pollution control laws. But extremes of drought and flood plus mounting pollution problems in the postwar period emphasized the shortcomings of the limited approach to basin management, and prompted the far-reaching legislation of 1963, which created the River Authorities.

Historical Background. Stream deterioration really began in England early in the nineteenth century. Industrialization itself produced massive amounts of wastes and gave rise to an explosive increase in the size of the cities. At about the same time, the water carriage system of sewage disposal was introduced; wastes were no longer disposed of in the streets but carried into the rivers. This, combined with direct discharges to the rivers by industrial establishments, meant that large amounts of untreated municipal and industrial waste began to find their way into the streams.

By the middle of the nineteenth century, pollution had become a serious public health problem, particularly in such densely populated areas as Lancashire, Yorkshire, the Midlands, and London. Epidemics, destruction of fish life, and grossly offensive river conditions prompted the appointment of Royal Commissions. The Act contained strong provisions forbidding the discharge of solid and liquid wastes, but it also stipulated that no action could be taken unless the local government board was satisfied that "no material injury will be inflicted by such proceedings on the interests of such industry." This and certain other provisions hindered enforcement, and in the course of time it became clear that the Act was not being successfully implemented. River agencies of various kinds were then set up in some of the most heavily developed watersheds to administer the 1876 Act.

After World War II, a feeling emerged that previous legislation had in large measure been a failure. The situation has been described by Craine as follows: "The administration of regulatory law prior to the end of the Second World War tended to favor established uses giving dominance to the doctrine of protecting individual rights unless the exercise of such rights implied a flagrant violation of the public interest. With the war and throughout the immediate post-war period, a rapid shift in public attitude gave greater weight to the collective interest in the quality of the natural waters of the nation." Parliament passed the River Boards Act, which established thirty-two river boards-one in each basin in England and Wales and conferred on the boards, or transferred to them, functions relating to land drainage, fisheries, and water quality, and recognized them as local authorities so they could prosecute under the 1876 Act. It also made it an offense for any person to create or substantially alter an outlet for the discharge of industrial or municipal effluent to a stream or to make any increased discharge of industrial or municipal effluent without the consent of the river authority. It further provided that such consent was not to be withheld unreasonably. This licensing authority was the primary new feature of the act. The act also permitted the River Boards to enact by-laws to restrict existing discharges, but the boards were unable to achieve agreement with the Ministry of Housing and Local Government which had general review authority on this matter.

One small step in this direction was a provision enabling the Minister of Housing and Local Government to set up, where necessary, a regional sewage board for centralizing the treatment of municipal and industrial waste. In the course of time, one such board was set up in the Mersey area. This board replaced fifteen obsolete waste treatment facilities with a central plant, which also serves a large number of industries that formerly discharged their wastes directly to the Irwell River.

Present Organizational Arrangements. The *Water Resources Act* had its immediate origins in a series of three reports issued by a subcommittee of the Central Advisory Water Committee. The first report was a general assessment of the water situation. The second, which followed a severe drought during the crop season, found the potential irrigation demand sufficient to establish a need to control withdrawals from surface waters. The third and final report, the so-called Proudman Committee Report, set forth in some detail proposals for comprehensive water policy and unified administration of water use and development activities in river basin areas. It has very little additional to say about water quality management, but it is expected that the regulatory powers of the *Rivers (Prevention of Pollution) Acts of 1951* and 1961 will be more vigorously exercised by the new River Authorities which have broader responsibilities for water management than their predecessors, the River Boards.

Among the regulatory powers given to the River Authorities the following are particularly significant: authority to operate a comprehensive system for licensing water withdrawals from surface and groundwater sources; authority to introduce "charging schemes" under which water withdrawers will be charged on the basis of the quantity they have been authorized to withdraw, with some adjustment for source of supply, season of the year, uses to which water will be put, and the way water is disposed of after use; and authority to license *all* discharges of wastewater to underground strata as well as to streams.

These regulatory provisions in the Water Resources Act are associated with extensive multipurpose development powers which now reside in the River Authorities. The functions and power to carry out development works relating to land drainage, flood damage reduction, and fisheries were transferred from the River Boards, and, in addition, the River Authorities were given new powers to construct, operate, and finance multipurpose facilities, including water storage for municipal, industrial, and agricultural uses. The Act further strengthens River Authorities by giving them a key role in the collection and analysis of basic data and information and in the formulation of specific development proposals.

Of particular interest for our purposes is the provision that each River Authority is required to determine a minimum acceptable flow at each of the designated and agreed upon control points. "In determining the flow to be specified the River Authority shall have regard to the character of the inland water and its surroundings (and, in particular, any natural beauty which the inland water and its surroundings may possess) and the flow so specified shall not be less than the minimum which in the opinion of the River Authority is needed for safeguarding the public health and for meeting (in respect both of quantity and quality) the requirements of existing lawful uses whether for agriculture, industry, water supply or other purposes, and the requirements of land drainage, navigation and fisheries."

At the national level, the Ministry of Housing and Local Government and the Ministry of Agriculture have long been the dominant cabinet agencies concerned with water. Under the new law, they continue to play a major role in their respective areas of concern. In addition, a new Ministry of Land and Natural Resources has been created to give supervision to data collection, surveys, and research related to water resources. A fourth agency of major significance is a newly established Water Resources Board, located administratively under the Minister of Land and Natural Resources, but with wide-ranging water policy, planning and advisory authority, particularly with regard to inter-basin transfers of water. The Board also has specific approval and directive authority over river authorities in performing their functions regarding hydrologic measurement schemes, minimum acceptable flows, water quality control and proposals for action.

Craine has commented as follows about the over-all situation: "River Authorities, however, are central to the new British scheme.

The law provides that their managing boards will be composed of representatives of local governments and of the central government in such proportions as to always assure local governments a bare majority and no more. It is in these 27 semi-independent authorities that the grass-roots issues of water management are faced."

In France, as in Britain, regional water management agencies are creatures of national law. The responsibility for formulating standards for water quality control and supervising their application was scattered among a number of agencies, both national and local.

Historical Background. The central authority for industrial waste control appears to derive from the Dangerous, Objectionable, and Unhealthy Establishments Act of December and its various implementing decrees. The Act placed all industrial and commercial establishments endangering the safety, health, or amenities of the community under the authority of the Prefect (chief administrator of a department, or administrative district). All these establishments are inspected by civil servants known as Inspectors of Classified Establishments. They operate under the instructions of the Prefect, and their recommendations usually carry considerable weight. Emphasis seems to have been put upon negotiations and advice to the industries on a case by case basis. Should negotiations fail, there are legal arrangements to compel compliance. This approach has not prevented the development of a severe industrial waste disposal problem.

Other water quality control laws relate to fish life, protection of drinking water, and protection of public and private watercourses. Under the Rural Code, any discharge harmful to fish life, reproduction, and nutritive value is punishable by a fine of 50,000-500,000 francs and/or imprisonment for ten days to one year. Water and forestry officials are responsible for enforcement, and because the penalties are heavy, they have tended to enforce the law with considerable restraint. The Public Health Code, in addition to some more specific provisions, provides penalties for any person who allows substances injurious to health to enter the public drinking water supply.

The engineers of the Navigation Service enforce the water quality control regulations pertaining to public (navigable) watercourses. Authorization must be obtained for any discharges into these watercourses. Infractions are dealt with by the administrative courts.

On so-called private (non-navigable) watercourses, the Civil Code stipulates that a riparian owner has the right to use the water flowing along or through his property, but not to impair its quality it must leave his land in its initial condition. As in other countries having riparian law, this provision is ignored more often than it is observed. Additional provisions of the Rural Code entrust the Rural Engineering Service with water quality control functions on the private watercourses.

Evaluations of pollution control in France vary. Still, the preponderance of evidence seems to be that it has been rather ineffective. The French government indicated that a large number of industrial concerns were discharging harmful effluents. The breakdown by watershed was estimated as follows: Seine, 1,000; Loire, 600; Garrone, 300; Rhone, 750; Rhine-Moselle, 500; Adour, 75; northern rivers, 350; watercourses flowing to the Brittany and Normandy coasts, 400; Atlantic Coast watercourses, 100; Mediterranean watercourses. 150. As of 1965, only 3 million people out of 25 million people in towns and cities with populations of over 2,000 were provided with any treatment facilities.

Discontent with the previous record of pollution control and the fragmentation of responsibility played a large role in the creation of a Commission on Water, attached to the Commissariate General du Plan. As the result of the work of this Commission, a comprehensive new law was passed by Parliament.

Present Organizational Arrangements. The law might be described as revolutionary. While France has had less experience with regional approaches to water quality management than any of the other countries considered in this study, the new law foresees going beyond what has been done elsewhere, particularly in the area of using economic incentives for water quality management.

Three features of the law are of particular importance: (1) it established agencies in each of the six river basins of France to serve as regional branches of the national government; (2) it gives these agencies powers to implement regional programs of water quality management; and (3) it places primary emphasis on charges, particularly effluent charges, to finance the program and to co-ordinate private and local government waste discharge decisions with the objectives of the regional agency. In other words, it views charges levied on effluents as the primary means for inducing private and local governmental control of waste discharges. The act does not repeal previous water quality control legislation, but it anticipates that most of it will be displaced as the provisions of the new law become effective.

The act is extremely general, indeed it is what the French term a skeleton enactment. This means that Parliament has defined broad principles, and the administration is to state the provisions of the act in the form of rules (decrees in the Council of State). Very few such decrees had been published, and consequently much of what the law intends must be inferred from the statements of those who helped frame it. The Act takes a very broad view of the water quality problem as involving "any occurrence liable to cause or increase the pollution of water by altering the physical, chemical, biological, and bacteriological problems of surface water, groundwater or seawater within territorial limits." It also sees the water quality problem within the context of over-all water resources management and gives the basin agencies general water resources management powers.

The generalities of the act have given rise to some disagreements over its proper interpretation with respect of collective facilities. At first, some well-informed commentators (Levy-Lambert, for example) thought that it permitted the agencies to build and operate collective facilities. However, it is now agreed that the basin agencies must act primarily through grants, loans, and contracts with private and public bodies. It is foreseen that these will be used in such a way as to achieve scale economies and lead toward an economically efficient solution. A new law will be needed to enable the basin agencies themselves to build and operate water control quality measures. Since it is now widely agreed that the new basin agencies should have this power, no difficulty is foreseen in getting such a law enacted by Parliament.

The activities of the river basin agencies will be supervised and co-ordinated by the central government. Each management area will be mapped out by the standing interministerial committee on development planning problems, and a national water committee under the direct authority of the Premier will advise on all water resource development and distribution projects that are of major importance or that affect several basins. A River Basin Committee will be set up for each basin or, where appropriate, for a group of basins. The water users (including waste dischargers), local communities, and the administration will be equally represented on it. The committees are to advise upon the desirability of any works or alterations contemplated in the area for the common benefit, but programs for water management (water quality improvement, for instance) must be approved by the central government before they are executed.

The basin agencies will be allowed to levy charges on corporations and individuals "taking into account the extent to which they have made the provision of facilities useful or necessary or will benefit therefrom." While the 1964 Act sees charges as a way of financing largescale facilities such as reservoirs, treatment facilities, and measures to improve stream assimilative capacity, it also views them as a means of compelling waste dischargers to compensate for the "external" costs they impose on the general economy. The original plan of setting up organizations only in areas where there seemed to be particular problems was abandoned, and basin agencies were established throughout the country, each encompassing a whole hydrologic area. It was felt that interdependence between the users of the basin from the source to the mouth could no longer be neglected; at the same time it was recognized that some geographical decentralization might be desirable in the larger basins, and the matter of sub-basin authorities was left open. Such authorities, if set up, would have considerable freedom of action, but would remain under the general authority of the agency in charge of the whole basin.

The regional approach to water quality management in France appears to have been influenced by conceptual matters such as hydrologic interdependency and external costs, and also by developments in other countries. There are many references to the work of the English River Boards in the French literature of the early sixties, and particular attention was given to the developments in the Ruhr area. The French agencies are in many ways like the German water associations, but they differ in governmental structure and in the role assigned to effluent charges in the management systems.

Delaware River Basin Commission

Many of the problems in the Delaware Basin are associated with the great variability in flow of the river, combined with highly developed flood plains and heavy demands for water for municipal and industrial use and waste disposal. On August 19, 1955, the date of the valley's worst flood of record, the peak discharge of the river at Trenton was about 330,000 cubic feet per second. During the persistent drought of the early and middle sixties flows were well under 1,500 cubic feet per second at Trenton every summer and fall, despite reservoir releases to increase flows. The main water quality problems are in the Lehigh Valley and along the Delaware estuary. The latter is far more important in terms of the quantity of water and the number of people and economic activities affected, and much of the succeeding discussion applies particularly to the estuary.

Historical Background. By the late seventeenth century, waterpower was being developed at falls along the tributaries of the Delaware River, and water-powered mills had made the Brandywine watershed (tributary to the Delaware estuary) a center for the production of flour, gunpowder, textiles, and paper. A combination of natural resources, including plentiful water, navigation opportunities, and advanced cultural and political development enabled the Delaware area, especially the estuary and the Lehigh watershed, to grow rapidly and hold its position as the center of American manufacturing for many years. Despite the growth in population and the magnitude and nature of industrial development, very little was done to deal with water quality problems until the last few decades. Even in 1940 little treatment of any kind was given to wastewaters in the region.

The idea of setting up an interstate compact for comprehensive development of the basin's water resources was first suggested in the 1920's when New York City wanted to withdraw 600 million gallons of water a day from the upper Delaware Basin. However, the New York proposal led not to a compact but to a Supreme Court case and a decision in 1931 that permitted a diversion of up to 440 million gallons a day, and authorized a development of the upper Delaware designed for New York City's water supply with incidental low-flow augmentation to reduce claimed downstream damages in the basin itself.

The Interstate Commission on the Delaware River Basin (INCODEL) was formed. The Commission was not an interstate compact agency as such but an agency created by parallel legislation enacted by Delaware, New Jersey, New York, and Pennsylvania, and operated from the beginning on the principle of voluntary interstate cooperation.

INCODEL had two main committees: one to deal with water quantity and one with water quality. The Committee on Water Quality, whose members represented the departments of health in the four states, came to dominate the activities of INCODEL. The committee's orientation was not toward comprehensive planning of the water resources of the basin but toward dealing with immediate water quality problems. For this purpose, the committee prepared a reciprocal agreement on pollution control, which classified the waters of the river and its tributaries into four categories or zones based on use, and prescribed standards of treatment for each. The agreement was promptly adopted by the department of health in each of the four states. The committee also drafted a uniform pollution control law which was passed in all states except Pennsylvania by 1941. The strong opposition in the Pennsylvania legislature reflected both the atttiude of industry and the fact that the City of Philadelphia accounted for a substantial portion of the untreated sewage discharged into the Delaware.

When INCODEL offered its plan again, the measure was passed by the Pennsylvania legislature, and construction of waste treatment plants by municipalities commenced immediately after the war. This was counted as a major victory by INCODEL, and the agency then appears to have rested on its laurels. The importance of waste treatment plant construction is by no means to be minimized, but on the average not much more than primary treatment was provided, and the residual municipal wastes together with industrial discharges from a growing economy continued to place heavy oxygen demands on the stream.

For a number of years INCODEL did comparatively little in the way of multipurpose planning. It displayed a preference for parallel state legislation, and not until the late forties and early fifties did INCODEL attempt to initiate broader planning in the basin. Engineering firms were engaged to survey the situation, and they presented a plan for the states to join in a compact for the construction of a number of reservoirs. Pennsylvania once more proved to be the critical state. Upon the negative recommendation of a special study committee, which decided that the benefits to Pennsylvania were not worth the costs, the Pennsylvania legislature did not pass the plan.

While the Pennsylvania study was going on, New York filed a petition with the Supreme Court in 1991 for modification of the 1971 decree to permit diversion of 800 million gallons per day. The Supreme Court appointed a special master to conduct hearings among the litigants and to make recommendations to the Court. Agreement was finally reached on a plan which would permit New York to divert 800 million gallons per day with compensating releases during low-flow periods from reservoirs, which New York would construct.

The Supreme Court without opinion adopted a decree as agreed upon among the party states and recommended by the master. By this time, the INCODEL proposal was virtually dead. Two severe hurricanes struck the Delaware Basin causing floods which took about 100 lives and caused some $125 million of damage. This set in motion a study by the U.S. Corps of Engineers. Even before the flood and the initiation of the Corps of Engineers planning study, the governors of the four basin states and the mayors of New York and Philadelphia had organized the Delaware River Basin Advisory Committee and given it responsibility for assessing water problems in the basin. The Committee, working with leading citizens of the basin, stimulated the formation of a private nonprofit corporation, Delaware River Basin Research, Inc., which financed a study of governmental organization for water resources administration at Syracuse University. Upon completion of the study in late 1989, the Advisory Committee drafted a compact to establish a regional water management agency, which would be an agency of the four states and the federal government. The states and federal agencies acted quickly, and in 1991 Congress approved the federal-interstate compact establishing the Delaware River Basin Commission.

Before turning to an actual description of the compact and its provisions, it should be noted that, when the Advisory Committee was drafting the compact and the Corps of Engineers was preparing its plan for the basin, municipal and industrial water supply, flood damage reduction, and water-based recreation were regarded as the main considerations, although salinity and other water quality problems commanded some attention. It now appears to the staff of the Commission and to the Water Resources Association of the Delaware River Basin (a nonprofit corporation founded in 1989 to represent a number of citizens' associations) that the water quality problem is the most serious one in the basin.

Another necessary background item is a brief history of federal laws concerning water quality. The successful completion of the INCODEL program of providing basic (at least primary) treatment like similar endeavors by ORSANCO in the Ohio Valley and by a number of important states largely preceded any effective national legislation in water quality matters. Also, the development of the Delaware River Basin Compact was independent of federal initiative in water quality matters. Nevertheless, the national scene has been characterized by vigorous law-making activity in recent years and it is within the resulting legal context that regional agencies must, at least for the time being, operate.

The federal government took its first step into what had been an exclusive area of state sovereignty. The details of the act need not detain us. It primarily supplemented state and interstate efforts with technical aid and services. Federal enforcement action was authorized, but only in interstate situations and with the consent of the states in which waste discharges causing water quality degradation were deemed to originate. The act was rewritten. The new act strengthened federal enforcement authority in connection with interstate water quality problems and introduced grants-in-aid to subsidize municipal waste treatment plant construction, although the amount authorized was comparatively small ($50 million a year) and the limit for individual projects relatively low ($600,000). The law also aimed to encourage state and interstate efforts by various means primarily subsidies for administration and it initiated a program of comprehensive water quality surveys.

The Water Pollution Control Act was amended again in 1991. Enforcement authority was broadened to include all "navigable" waters, and subsidies for the construction of municipal waste treatment plants were increased. The 1991 Act also provided for the inclusion of storage in federal multipurpose reservoirs to augment low flows for water quality improvement. In general, the costs of such storage were to be considered nonreimbursable.

The law was once again amended in 1995. The *Water Quality Act* was highly significant because it broadened federal jurisdiction through a provision requiring the establishment of standards of quality for all interstate waters. The states were required to set such standards (to be approved by the Secretary of Health, Education, and Welfare) by June 30, 1967, or face the imposition of federal standards. Federal subsidies for municipal waste treatment plant construction were once more increased. This represented an effort to broaden and improve the status of the federal water quality effort. The FWPCA was subsequently moved from the Department of Health, Education, and Welfare to the Department of the Interior. The intention was to broaden the presumably narrow public health view of water quality control. The most recent law is the *Clean Water Restoration Act*. The feature of primary interest to us is a further large increase in the federal subsidies available for municipal waste treatment facilities. It authorizes $3.4 billion for construction grants for the fiscal years. The former limitation on individual grants is modified. If certain conditions are favorably met, the federal share can now rise to as high as 55 per cent of capital costs. To provide incentives for developing comprehensive water quality planning for entire watersheds, the act offers assistance in the financing of such undertakings up to 50 per cent of the cost of the expenses of a planning agency for a period of three years if it is requested by the relevant governors and certain conditions for representation are met. The central elements of current federal law of interest here are: (1) heavy subsidization of waste treatment plant construction; (2) technical and planning assistance to state and interstate agencies; and (3) substantial enforcement powers which have as yet been comparatively little used.

Present Organizational Arrangements. The Delaware River Basin Commission is the only interstate-federal compact agency in the United States and the only regional agency to be given such comprehensive powers. The central objective was to create a river basin commission that would devise and administer a comprehensive multipurpose water resources plan which would "bring the greatest benefits and produce the most efficient service in the public welfare."

The compact states that the United States "consents to, and joins" an intergovernmental compact creating the Delaware River Basin Commission which is to be "an agency, an instrumentality of the governments of the parties." The Commission has five members, one from each of the four member states and one from the federal government. The commissioners are the governors of the four basin states and the Secretary of the Interior representing the President of the United States. Each commissioner may specify a voting alternate. In each case, these have been men knowledgeable in the field of water management.

The Commission is charged with developing a "comprehensive plan for the immediate and long-range development and uses of water resources of the basin" and a shorter-range "water resources program" based on the comprehensive plan. The Commission is granted the power to plan and implement the development of water resources by acquiring or building and operating and maintaining dams, reservoirs, and similar facilities to control and develop the water supply; to regulate the flows so as to regulate water quality; to control and abate existing pollution; to provide for flood protection; to institute sound practices of watershed management including prevention of soil erosion and promotion of land reclamation and sound forestry practices and fishing and wildlife conservation measures; to promote recreational uses of the river; to develop and operate facilities for the generation and transmission (but not for direct distribution to consumers) of hydroelectric power and to set rates and charges for such power; and to regulate and control withdrawals and diversions from the waters of the basin.

The Commission also has a wide variety of powers for implementing the plan including the authority to delegate certain of its powers to instrumentalities of signatory governments, to establish cost-sharing standards and formulas for the apportionment of costs among the different purposes included in multipurpose programs for the sharing of financial responsibility with the signatory parties, public bodies, groups, and private enterprise. The compact authorizes broad borrowing powers and the power to issue bonds. The parties agree to provide capital funds for the projects of the Commission in accordance with the cost-sharing provisions to be established by it. The compact provides for the apportionment of the annual expense budget among the signatories with all of them undertaking to include their portion in their respective budgets "subject to such review and approval as may be required by the respective budgetary processes." The Commission has the power of eminent domain and the state signatories agree not to undertake any development projects in relation to the Delaware River Basin unless the project is approved by the Commission and is in accordance with the comprehensive program. The federal government agrees to substantially the same terms.

The compact gives the Commission extremely broad power for the direct implementation of various measures affecting quantity and quality of water in the basin, both surface and ground. It is intended that the Commission shall exercise such powers only if another federal or state agency cannot or does not provide the same functions efficiently in accordance with the comprehensive plan.

Specifically with regard to water quality management, the Commission is given the powers to undertake investigations and surveys and acquire, construct, operate, and maintain projects and facilities to control potentially adverse waste discharges and abate or dilute existing waste discharges affecting quality of the water resources of the basin. It may invoke, as complainant against waste dischargers, the power and jurisdiction of water quality control agencies of the signatory parties. The Commission may assume jurisdiction to control future waste discharges whenever it determines after investigation and public hearings that the comprehensive plan requires it. The Commission, after public hearing, may also classify the waters of the basin and establish standards for treatment of municipal, industrial, or other wastes. It also has the power, after hearings, to amend and repeal rules, regulations and standards. The signatory parties also pledge themselves to pass whatever legislation may be necessary to control wastes in accordance with the comprehensive plan. The Commission also has direct enforcement powers and may, after investigation and hearing, issue orders to any persons or person or public or private corporation to cease any discharge which it determines to be in violation of the rules and regulations that have been adopted. The orders are subject to appeal in any court of competent jurisdiction.

To summarize, the Commission has far-reaching powers to plan, design, construct, operate, and finance facilities for water quality management if it can demonstrate that these are effective and efficient means for achieving the water quality objective and that the objective could not be so efficiently achieved by other units of government. The Commission also appears to have considerable scope under federal law to proceed as it deems best in the matter of standard-setting and control of discharges.

Comparative Discussion of the Management Approaches

We suggested that if regional agencies were to be able to achieve economic efficiency they should have the power to: analyze and implement a wide range of alternatives; influence the pattern of land use as it relates to the water management program; and articulate private and local government decisions with their own operations through regulation or charges. We now look at the four cases outlined above in relation to these powers and in relation to the nature of the representation of, and the opportunity to influence decisions by, the affected parties.

Ability to Evaluate and Implement a Wide Range of Alternatives. A regional agency is in a much better position to improve the efficiency of its water quality management system when it can make use of large-scale measures such as streamflow regulation, mechanical reaeration of streams, direct treatment of streams via oxidation lakes and other means, collection and treatment of municipal and industrial wastes in central plants serving entire regions, piping of wastes away from critical reaches of streams, and groundwater recharge. These measures will probably be most beneficial in highly developed basins, but they should be considered in others as well. In general, the regional approaches discussed in this chapter are cognizant of such opportunities, and the laws governing them make provision for their implementation. The only exception is England where the law establishing the River Authorities, the documents leading up to it, and discussions following it, all suggest that the River Authorities are not expected to undertake activities of this kind.

The *Genossenschaften* are a special case. They began with the idea of establishing integrated basin-wide systems and they have had a number of decades to implement them. The range of alternatives evaluated and implemented is wide, and virtually no possibility is ignored. Nevertheless, two factors may militate against the actual identification and implementation of an optimum system in the Ruhr. One is what appears to be an excessive separation of water quality and water quantity considerations. The other is that little or no use is made of formal systems analysis, despite the broad scope and responsibilities of the organizations and the utility of this mode of analysis for both planning and operating water quality management systems. This reflects a more general lag in the application of systems analysis to water resource engineering and economics in Europe. To what degree such techniques as computer simulation and more sophisticated analysis of hydrology might improve the systems must remain an open question. Limited information from U.S. research and practice suggests that these are important planning, design, and operating tools.

British practice has tended to place virtually complete reliance upon treatment at individual points of waste outfall. Efforts toward achieving scale economies in waste disposal have been few. *Drainage of Trade Premises Act* authorized the discharge of trade wastes into public sewers subject to the approval of the local authority, and the *Rivers (Prevention of Pollution) Act of 1951* encouraged the integration of separate local sewer systems and treatment plants. Up to this time, however, it cannot be said that there has been any systematic exploitation of the opportunities provided by collection and treatment of wastes on a regional scale (with the exception of some efforts by the Thames Conservancy), not to mention the array of other possible measures that might be implemented on a river basin basis. Even the new river agencies have no authority to implement a water quality management facility as such, although they can construct reservoirs and operate them for streamflow regulation.

The probable economies of broader regional scale measures has not gone unrecognized by many English water quality professionals, but neither the law nor the various available interpretations of it suggest that much emphasis will be placed on direct investment and operation of facilities for water quality improvement by any of the River Authorities. This contrasts strikingly with the emphasis placed upon the river basin development and management approach to water supply and the strong role assigned to the River Authorities. The situation probably reflects the long history of water quality control based on efforts to impose restrictions on individual waste dischargers, and the fact that several laws had not yet been fully tested at the time of the studies and debate on the River Authorities' legislation.

The French agencies, unlike some of the other regional agencies discussed in this report, seem to have an absolutely clear objective economic efficiency in the full sense of the term. The aim is to establish a system in which the incremental costs of further improvement of water quality balance the incremental benefits, and in which the full range of alternative ways of improving water quality can be assessed and all measures brought into optimal balance. Current legislation does not give the river basin agencies authority for direct investment in regionalscale works, but they can make grants or negotiate contracts with private or public bodies to construct the facilities needed for an economically optimal system. There seems to be considerable sentiment to supplement the basic law by giving the basin agencies the authority to build and operate control measures.

The Delaware River Basin Commission has authority under the compact to plan, construct, operate, and finance measures for water quality improvement, but only if it can do so more efficiently than other agencies. The Commission seems particularly well-suited to undertake such measures although it would need a much larger staff for planning, economic evaluation, and program development. The Federal Water Pollution Control Administration study of the Delaware estuary went far in identifying and evaluating alternatives in terms of both combinations of measures to improve water quality and the objectives to be met. Although the study did not explore all the possibilities for realizing regional scale economies, its findings seem to justify direct action by the Commission in the construction and operation of works if it is to achieve the most efficient system for dealing with the water quality problems in the estuary. Moreover, in the matter of systems analysis tools the Commission is far ahead of any comparable regional agency in the United States or abroad. A board of consultants has strongly urged the Commission to use these tools and to proceed to implement a program of regional measures.

As each of the countries discussed makes at least some use of subsidies, it is worth pointing out again that financial aid of this kind-whether for streamflow regulation, municipal waste treatment plant construction, or industrial waste treatment facilities may make an inefficient facility seem financially attractive and thereby distort the planning and implementation of alternatives. This suggests that a policy of subsidization would be more effective if the grants were not restricted to particular facilities but could be used by regional agencies where needed to implement an over-all regional plan for water quality improvement.

Ability to Integrate Related Water and Land Uses. As a rule there is a strong tie between considerations for water quantity and water quality. Streamflow regulation through reservoir storage not only influences the amount of water available for diversion, depletion, and in-stream uses but also affects the quality of water in the stream. Therefore, quality considerations should be taken into account in the planning, design, and operation of reservoirs. Furthermore, restrictions on water usage such as permits or charges for water supply withdrawal influence water quality. If an industrial enterprise is induced to recirculate its water, the waste load is more than likely to be reduced through internal treatment or waste recovery processes. Moreover, the residual waste load will be in a smaller volume and therefore less costly to treat.

Location patterns may greatly influence the costs associated with waste disposal along an estuary. For example, the establishment of a new industrial discharge immediately above a public water supply intake may impose a large enough burden to make it worthwhile for society to consider changing the pattern of development. Institutions for land and water management should be such that these interrelationships will be evaluated and properly taken into account.

In England, the Ruhr, and the Delaware Basin the interrelationships between water quantity and quality have been taken into account by putting them both under the authority of one agency. In France a separate agency will probably continue to be in charge of reservoir construction and operation. The law implies that releases for water quality improvement will be arranged on a reimbursable basis by the water quality management agencies and that they will participate in reservoir planning. It may be that an interagency relationship of this kind can achieve the same results as a single agency with the authority to internalize all direct interdependencies.

Because of the wide scope of their authority over water resources, the *Genossenschaften* are in a very good position to integrate all aspects of water management. The fact that institutional development has created a somewhat artificial distinction between water quality and water supply, may be of minor importance because the opportunities for flow regulation to improve water quality are so limited in the Ruhr. The Ruhrverband's charges for water withdrawal have caused industries to develop internal water recirculation systems, and the result has been a reduction not only in water intake but also in the volume of wastewater and the cost of treatment.

A certain amount of co-ordination with land-use development has been achieved by the staff of the Ruhrverband working with the Siedlungsverband, which is charged with general land-use planning in the industrial region of North Rhine-Westphalia. In devising its over-all plan for the region, the Siedlungsverband considers the impacts of land use upon water management. This is the only case among those considered here where there is a reasonably comprehensive and systematic co-ordination of land-use planning and water resources planning.

The English River Authorities have several potentially effective management controls at their disposal. They have the authority to regulate the withdrawal of water from natural watercourses, to regulate discharges of wastewater, to charge for the withdrawal of water, and to construct and operate water management facilities with primary emphasis on reservoirs for storage and flow regulation. While charging schemes are also possible, there seems to be a strong emphasis on direct controls, which, as was pointed out in earlier chapters, are often inferior to pricelike devices as a means of achieving an economic optimum. Nevertheless, a number of critical interdependencies are clearly recognized and tools are provided to influence them. There is also some recognition of the influence of land use and development upon water quality management. Most of the River Boards had arrangements with local planning authorities for all planning applications involving new discharges of trade or sewage effluents to be referred to the River Board for review and advice. The River Authorities are expected to make similar arrangements. This system provides some protection against gross mistakes in the location of activities.

In sum, while most of the critical interdependencies appear to have been recognized in providing management tools to the English River Authorities, the heavy dependence on direct controls and failure to integrate water and land planning must be counted as deficiencies. In principle, the French law provides for thoroughgoing management of both water quality and water quantity. The new law contemplates that charges will be levied upon water that is diverted and lost from the system, as well as upon effluent discharges. It is envisaged that two related but distinct fees will be paid by those who withdraw water and return waste to the system. It is also foreseen that the basin agencies will use grants, contracts, and subsidies to bring all elements in the water system into optimal balance. If, as Is expected, the agencies are given the authority for direct investment, this power will be exercised in the same context. While these are only intentions at this point, it is fair to say that interdependencies between water quality and water quantity have been clearly and explicitly recognized.

On the matter of interrelationships between land use and the cost of water supply or waste disposal, the French law is largely silent. The same is true of various discussions of the law. The act does not go beyond providing for the setting up of protective zones around sources of public water supply.

The Delaware River Compact gives clear recognition to interdependencies between water quantity and quality, and in principle there is no reason why the Commission's authority to set standards, levy charges on effluents and on water withdrawals, and construct and operate facilities should not permit it to take optimal account of all interrelations between various water uses including waste disposal It is somewhat doubtful whether this will actually be accomplished.

The matter of associated land use came tip during the framing of the compact, but no action was taken on suggestions that the proposed commission be given open-ended authority and jurisdiction permitting other functions to be added later by the states. Apparently it was thought that opening the door to other government functions would bring forth complexities and political difficulties that would defeat the whole enterprise. As ii result the compact is silent oil the matter of relationships with land use. The Delaware Commission might have some influence on land-use patterns through its stream classification and charging powers, but in general the issue has not been confronted. Since location decisions should not turn upon water quality considerations alone, it may be that no appropriate consideration of this variable can occur until the basin has a regional land-use organization with regulatory as well as planning authority.

In summary, in each of the cases considered, interdependencies between water quantity and quality are explicitly recognized, and tools are provided for managing these interrelationships. Particular types of tools used or contemplated vary; some agencies place heavy emphasis on direct controls whereas others rely more heavily on indirect measures such as charges on water withdrawals and or waste discharges. The only instance of reasonably systematic recognition of the interrelationships between emerging land-use patterns and water quality management problems is in the Ruhr. It may be significant that this is also the only region with an effective regional land-use planning agency.

Ability to Articulate Private and Local Government Decisions so as to Achieve an Efficient System. The extent to which co-ordinated action on a regional basis is achieved through direct planning, design, and operation of structural works by the regional authority will vary from case to case. In some instances, such as the operation of a reservoir or an in-stream treatment facility, it is in the interest of efficiency for the regional agency to provide and operate the facility because no other agency has the responsibility or capacity to do so. Major efficiency gains may also result from central operation of treatment plants at individual industrial and municipal outfalls. In the Ruhr area, the *Genossenschaften* operate all conventional treatment plants. Industries have frequently incorporated internal waste recovery and pretreatment processes, but it is one testimony to the efficiency of the *Genossenschaft* system that municipalities and industries have chosen to pay effluent charges rather than undertake treatment of their wastes.

In few cases, if any, could regional waste disposal considerations justify the direct design and operation of industrial processes by a regional agency, although internal process design, internal materials recovery, and by-product production can greatly reduce industrial waste discharges. The role of a regional agency should be to devise some method of causing industrial waste disposers to take account of the costs they impose on waste treatment facilities or successive users when they plan, design, and operate their production processes. Where the regional agency does not construct and operate treatment plants, it must use indirect means to induce optimal levels of waste load reduction at municipal and industrial outfalls.

Although the *Genossenschaften* do operate all conventional and some unconventional treatment devices, they still encourage industries to reduce their waste discharges. One incentive in this direction is the effluent charge which varies with the quality and quantity of waste discharged either to treatment plants or directly to the streams. The charge system, combined with technical and marketing assistance, has produced a substantial reduction in industrial waste loads. The effluent charges established in the Ruhr are not theoretically ideal, but they have had considerable practical effect in articulating the decisions of industrial waste disposers with the planning and operation of the regional works provided directly by the *Genossenschaften*.

The English authorities are equipped with a wide array of water management tools. It is foreseen that each case will be decided on its own, and each license will specify permitted waste discharge. Since there is little or no evidence that the English authorities will place much emphasis upon possibilities of exploiting economies of scale within their regions, it appears that these direct discharge controls will be the primary means for controlling water quality. In principle, it is possible to devise a system of direct controls tailored to conditions as affected by individual outfalls which would correspond to a regional economic optimum (considerations of regional scale economies aside). In practice, this technique requires more information and is less flexible than the effluent charges device. However, it may have the advantage of greater administrative ease. It does not, of course, provide a source of revenue for financing the activities of the regional agency, including the exploitation of economies of scale in water quality management.

The English law permits the River Authorities to devise and administer "charging schemes" under which "abstractors" will pay for water removed from surface or underground sources, and specifies that the rate may be adjusted to reflect differences in: (1) characteristics of the source of supply; (2) season of the year when water is withdrawn; (3) how the water is used; and (4) how it is disposed of after use. In principle, therefore, charges might be used as a flexible tool of management. The law might be interpreted to mean that the charge could be used to create an economic incentive for waste reduction activity by municipalities and industries and thus serve as a supplement to the administration of licenses for effluent discharges. At best, this would involve some problems. For example, some industries withdraw from the system of a water wholesaler but discharge directly to a stream. In this instance, it would be cumbersome for the River Authority, which can only levy charges on intakes, to devise a method that would provide the industry with an appropriate incentive to reduce its waste discharge. It must be said also that neither the law nor the associated literature suggests that charges be used as an incentive for waste control. The discussion is almost completely in terms of charges for withdrawals.

The French river basin agencies are expected to take cognizance of the economies inherent in regional action, and are authorized to promote the establishment of large-scale treatment plants for municipal and industrial waste, streamflow regulation, and other regional scale measures, when economic analysis reveals their desirability. Their water quality operations are to be financed by effluent charges, which are also intended to provide a way of inducing waste dischargers to take account of the costs they impose on the general economy when they make their waste disposal decisions. However, the law specifies that the rate schedules must be approved by the local river basin committees, and there is some doubt that the committees can be counted on to establish charges that actually reflect the external costs imposed on the general economy. Public and private users of the watercourse (municipalities, industries, recreational interests, etc.) have two-thirds of the seats on the committee and the central govenment has one-third. Nevertheless, it is intended that the effluent charges become the basic device for achieving control of effluents, particularly from industry. Moreover, the law specifies that all those affecting quality by their actions are subject to charges. Thus any action which reduces the waste assimilative capacity of the river-such as a deep impoundment might do, for example is subject to charge under the legislation.

How the Delaware Commission will proceed on these matters is still unclear. The Federal Water Pollution Control Administration study of the estuary found that minimizing the cost of treatment on a regional basis (even neglecting the possibility of largescale measures) would be much less costly than following the usual administrative procedure of requiring that all dischargers meet a single effluent standard. In other words, given the water quality goal in the estuary, it is much less costly to remove wastes to a higher degree at some outfalls than at others. The savings stem from a number of factors including scale economies in treatment methods and the dynamics of waste assimilation in the estuary related to the spatial pattern of discharges. Savings in cost of 50 to 100 per cent are possible in the estuary area if a cost-minimizing procedure is adopted. Another study found that effluent charges can be used to induce a pattern of treatment similar to that of the least costly treatment system, and that a charge of the necessary level (8¢ to 10¢ per lb. of oxygen-demanding material) was unlikely to disrupt the regional economy. The Panel of Consultants to the Delaware River Basin Commission has recommended that this approach be thoroughly explored. Whether it will prove politically feasible to implement the technique is still an open question.

Opportunity for Affected Parties to Influence Decisions. Achieving a supporting constituency and making sure that all relevant values are reflected in the decision-making process require some means of communication with the parties affected by the decisions. Knowledge of how the political power structure of institutional arrangements influences outcomes is too limited to permit linking it clearly to the economic efficiency objective. The matter of representation has nevertheless loomed large in the deliberations leading to the regional approaches considered in this chapter, and the laws establishing the regional agencies have all been quite specific on this point.

In drafting the plans for the Emschergenossenschaft, particular attention was accorded to the representation of interested parties because this was what had caused the failure of an earlier effort. The outcome was a political structure that seems highly unconventional from the perspective of the United States where it is unusual for private interests to be officially represented on public bodies with broad management authority.

Industrial interests were given a defined place in the assembly and on the board of directors of the Genossenschaften, and the Directors feel that this has given industry a strong sense of participation and reduced resistance to the provision of information and other co-operative services by industry. Communities are also represented in the assembly and on the board as a matter of legal right and are meant to speak for and promote the broader public interest. The direct representation accorded industry does not seem to have blocked attention to other values, and visitors to the Ruhr are impressed by the attention given to appearance and recreation opportunities. There is no way of knowing how the organizations would have fared if political power had been distributed in some other way, but the managers of the agencies believe that careful attention to the representation of affected parties has contributed to effective and efficient operation in the general interest of the area.

In England, the Proudman Committee recommended that the new river authorities follow the Ruhr pattern and have boards composed of representatives from user groups. However, a minority report favored continuation of the River Board pattern with members representing the local governments and the relevant ministries. The compromise which emerged specifies that the river authorities should range in size from twenty-one to thirty-one members, some to be appointed by local authorities and some by Ministers "because of their knowledge of particular aspects of the river authorities' work or interests affected by it." The various local taxing authorities are to be represented in accordance with the distribution of taxpayers in the area of each constituent local authority. The total number of members may vary, but local government representatives are to be a majority on each river authority.

To the extent that revenues are to be obtained from user charges, the specific users are not represented in accordance with financial contribution. However, local governments are answerable to the same clientele, and the water users and waste dischargers are represented through the local political process. Still, the distribution of power will be quite different than if industry were represented in some proportion to its water utilization or its role in the financing of the authorities.

The French public administrative establishments display still a third pattern, with half of the board members representing the central government, and the other half representing the local public and private users-communities, industry, agriculture, fishing, boating, etc. As the goal of the establishments is to achieve an economically efficient system, decisions will probably be guided by technical considerations at this level. The outcome may well be affected, however, by the actions of the advisory committees which are to be set up in each basin with water users, local communities and the administration each having one-third of the members. Charges are the critical financing and control measures in the French plan, but they cannot be levied until the rates are approved by the local committees. These committees may therefore play the central role in deciding how far-reaching water quality management actually will be. Since two-thirds of their members represent private and local government interests, the actual distribution of political power may not be unlike that of the Genossenschaften even though the organizational pattern is, at least superficially, quite different.

Representation takes a still different form on the Delaware River Basin Commission. The commission itself is composed of the state governors and their alternates and the Secretary of the U.S. Department of the Interior and his alternate. The alternates, at least so far, tend to be persons with a claim to technical expertise, and there is no direct political representation below the state level. The compact provides, however, that the Commission may appoint advisory committees "which may be comprised of representatives of the public and of federal, state, county, and municipal governments, water resource agencies, water using industries, water interest groups, labor, and agriculture." This is quite different from the situation in France where advisory groups are not only an integral part of the system, but they have certain critical powers of approval. Of all the regional agencies considered in this report, the Delaware Commission has the weakest links between itself and local governments and water users in the area. This may reflect the fact that the United States has three distinct levels of government whereas England and France have only two, and in Germany the federal government is not involved because all the Genossenschaften are within a single state.

The matter of an appropriate political structure for regional agencies is probably the least well understood aspect of water resources management. One of the most important issues seems to be the degree to which representation should reflect financial contributions. This has been a consideration in defining the structure of each of the regional agencies. However, in no case, except the Ruhr area, is there a one-to-one relationship between financial contributions and political representation. A separate but related issue is the matter of industry representation and its effect on the problems of implementing an effective water quality control program.

The Genossenschaften seem to feel that direct industry representation can ease the problems, and their experience suggests that this approach should not be lightly dismissed.

This discussion should not be taken to imply that the authors feel that industry has not been well represented in previous decisions concerning water quality control perhaps its views have been overrepresented. What we are suggesting is that responsible representations may be worth trying, i.e., representations in which industry activities would be brought into the open and its representatives required to make clear what they favor as well as what they are against.

Bibliography

ANDO, NISUKE *The law of pollution prevention in international rivers and lakes / Nisuke Ando.* The legal regime of international rivers and lakes = Le régime juridique des fleuves et des lacs internationaux / Ralph Zacklin, Lucius Caflisch (editors), 1981, pp. 331-370.

BETLEM, GERRIT *Legal barriers to liability for cross-border non-point source pollution / Gerrit Betlem.* Non-point source river pollution : the case of the River Meuse / ed. by Jan M. van Dunné, 1996, pp. 145-157.

BOYLE, ALAN E. *International law of environmental rights : remedies for pollution injuries / Alan E. Boyle.* Water in the Middle East : legal, political and commercial implications / Edited by J.A. Allan and Chibli Mallat, 1995, pp. 93-109.

CHAUDARY, V. (2000). Environmental Protection, Pointer Publishers, Jaipur.

COLLIARD, C.A. *Legal Aspects of Transfrontier Pollution of Fresh Water.* Legal aspects of transfrontier pollution / OECD ; [with contr. by Jean-Marc Bischoff ... et al.], 1977, pp. 263-282.

DICKSTEIN, H.L. *International Lake and River Pollution Control : Questions of Method.* Columbia Journal of Transnational Law, 1973, pp. 487-519.

DROZ, A. and H. SMETS.. *Transfrontier Pollution in International Water Basins.* Transfrontier pollution and the role of states. Paris: OECD, 1981, pp. 126-132.

DUNNE, JAN M. VAN *Transboundary Pollution and Liability Issues : Private Law vs Public International Law Approaches : the Cases of the Rivers Rhine and Meuse.* Acta Juridica, 1999, pp. 303-338.

ERICHSEN-BROWN, J.P. *Legal implications of boundary water pollution / J.P. Erichsen-Brown*. druk uit: Buffalo Law Review ; vol. 17 (1967). - P. 65-69. [S.l.: s.n.], 1967, pp. 5 p.

FETTIG, KRISTI *Criminal and Civil Remedies for Transboundary Water Pollution*. The Transnational Lawyer, 2002, pp. 117-154.

FLORIO, F. *Water Pollution and Related Principles of International Law*. The Canadian Yearbook of International Law, 1979, pp. 134-158.

FLORIO, FRANCO *Nota sull inquinamento delle acque non maritime nel diritto internazionale*. Rivista di diritto internazionale, 1963, pp. 588-605.

GURJAR, R.K. (1998). Water Resources, Environment and the People, Pointer Publishers, Jaipur.

HAASTEREN, J.A. VANBERG, R. VAN DEN (ed.) *Pesticides et eaux souterraines*. Conseil de l'Europe, 1993, pp. 57.

HOSETTI, B.B. (2002). Wetlands Conservations & Management, Pointer Publishers, Jaipur.

KLINE, JEROME NEIL *Intergovernmental relations in the control of water pollution / Jerome Neil Kline*. Natural resources lawyer, 1971, pp. 505-536.

LAMMERS, J.G. *Pollution of international watercourses : a search for substantive rules and principles of law / J.G. Lammers*. Nijhoff, 1984, pp. XXVI, 724 p.

MINGST, KAREN A. *Evaluating public and private approaches to international environmental disputes : statist and transnational solutions to acid rain pollution / Karen A. Mingst*. Natural Resources Journal, 1982, pp. 1-20.

MISHRA, S.P. (2004). Environmental Crisis, Pointer Publishers, Jaipur.

NEWSOM, GEORGE and J. GRAHAM SHERRATT.. *Water pollution*. Sheratt, 1972, pp. xii, 322.

PONTE IGLESIAS, Maria TeresaUniversidad de Santiago de compostela, *La contaminacion fluvial : cuestiones de responsabilidad internacional*. Universidad de Santiago de compostela, 1988, pp. 12.

POPESCU, D. *Water Pollution Control : a Legal Study*. Revue roumaine d'études internationales, 1971, pp. 173-179.

PRASAD, S.N. (2006). Global Environment Management, Aavishkar Publishers Distributors, Jaipur.

PRAT, J. and A. GIRAUD..Organisation for economic co-operation and development, *The pollution of water by detergents / [by] J. Prat [and] A. Giraud.* Organization for Economic Co-operation and Development, 1964, pp. 86 p.

RAWASHDEH, Z.A.N. (1996). Applied Hydrogeography, Pointer Publishers, Jaipur.

ROGGENBUCK, S. *Wasserverschmutzungen und internationales Privatrecht : zur privatrechtlichen Abwehr von Verschmutzungen internationaler Gewässer in der EG und in den USA.* Lang, 1997, pp. 308.

SATI, V.P. (2004). Water Resource Management and Policy Perspectives, Pointer Publishers, Jaipur.

SETTE-CAMARA, José *Pollution of international rivers.* Recueil des cours, 1984, pp. 117-217.

SHIVA, VANDANA *La guerre de l'eau : privatisation, pollution et profit.* Parangon, 2003, pp. 162.

SINGH, A. (1991). Environmental Degradation in Flood Prone Areas, Pointer Publishers, Jaipur.

SINHA, R.K. (2006). Development, Environment, Human Health and Sustainability, in 2 vols, Pointer Publishers, Jaipur.

STANDER, G.J. *Water Pollution : Problem of our Age.* South Africa International, 1970, pp. 89-96.

SYMONIDES, J. *International Legal Problems of the Fight against the Pollution of Rivers.* Polish Yearbook of International Law, 1974, pp. 99-114.

TECLAFF, LUDWIK A. and EILEEN TECLAFF.. *Transboundary ground water pollution : survey and trends in treaty law / Ludwik A. Teclaff, Eileen Teclaff.* Natural Resources Journal, 1979, pp. 629-667.

TECLAFF, LUDWIK A. and EILEEN TECLAFF.. *Transboundary toxic pollution and the drainage basin concept / Ludwik A. Teclaff, Eileen Teclaff.* Transboundary resources law / ed. by Albert E. Utton, 1987, pp. 27-50.

TRIVEDI, P.C. (2006). Industrial Pollution and its Management, Aavishkar Publishers Distributors, Jaipur.

TRIVEDI, P.C. (2006). Pollution Management, Pointer Publishers, Jaipur.

UTTON, A.E. *International water quality law / A.E. Utton.* Natural resources journal, 1973, pp. 282-314.

VIGHI, MARCO and CAROLINA SBRISCIA FIORETTI.. *Preface : the regulation of pesticides in Europe - past, present and future / Marco Vighi, Carolina Sbriscia Fioretti, Timothy Swanson.* Regulating chemical accumulation in the environment : the integration of toxicology and economics in environmental policy-making / ed. by Timothy Swanson and Marco Vighi, 1998, pp. xi-xx.

VISSER, F.M. *Internationaal Watertribuna.* Nederlands Juristenblad, 1983, pp. 1083-1087.

WILDE, K.T. deT.M.C. ASSER Instituut, *Riviervervuiling, in het bijzonder die van de Rijn, als probleem van volkenrecht.* T.M.C. Asser Instituut, 1971, pp. VI, 48 p.

WITASCHEK, F.V. *International Control of River Water Pollution.* Denver Journal of International Law and Policy, 1972, pp. 35-54.

Index